The Clusters Phenomenon in the Selected Central European Countries

The Clusters Phenomenon in the Selected Central European Countries

By

Magdalena Bialic-Davendra,
Drahomíra Pavelková and Eva Vejmělková

CAMBRIDGE
SCHOLARS
PUBLISHING

The Clusters Phenomenon in the Selected Central European Countries,
by Magdalena Bialic-Davendra, Drahomíra Pavelková and Eva Vejmělková

Reviewers: Jiri G. Drobny, MBA; prof. Ing. Karel Skokan, Ph.D.

This book first published 2014

Cambridge Scholars Publishing

12 Back Chapman Street, Newcastle upon Tyne, NE6 2XX, UK

British Library Cataloguing in Publication Data
A catalogue record for this book is available from the British Library

ISBN (10): 1-4438-5539-1, ISBN (13): 978-1-4438-5539-6

Magdalena Bialic-Davendra would like to dedicate this book to her beloved husband, and thank him for his infinite support in this journey.

TABLE OF CONTENTS

List of Figures

List of Tables

List of Maps

ACKNOWLEDGEMENTS

The authors are thankful to the Operational Programme Education for Competitiveness co-funded by the European Social Fund (ESF) and national budget of the Czech Republic, the grant No. CZ.1.07/2.3.00/20.0147 - "Human Resources Development in the field of Measurement and Management of Companies, Clusters and Regions Performance", which provided financial support for this research.

Further gratitude and thanks, for their dedicated time and co-operation, goes to:

- Petra Škodáková, Ph.D. (consultancy regarding benchmarking)
- Eva Maria Eckenhofer, Ph.D. (questionnaire survey in Austria)
- Ing. Lubor Homolka (co-operation on a statistical data evaluation)
- David Livingstone (proofreading)
- Scott Evan Andrews (translations)
- Dr. Donald David Davendra, Ph.D. (proofreading)
- Cluster managers and cluster representatives as well as representatives of institutions and agencies supporting clusters, who took part in the conducted research.

INTRODUCTION

Ever present in all spheres of our life, starting from countries and economies competing for leadership in the global context, up to individuals challenging one another, the phenomenon of competition constitutes the driving force of development. Competitiveness can be characterised by many layers. In the field of economy, it can be described by numerous theories and economic concepts; however defined, the most concise and genuine appears to be that of Porter's (2008) view describing competition as "one of the society's most powerful forces for making things better in many fields of human endeavour". Competitiveness is an indispensable element of the extremely complex and multi-dimensional phenomenon of clusters.

The cluster building process began during the 1990s in various European Union countries. The main reason was to strengthen industry's ability to innovate and increase national competitiveness. While the global economy became increasingly influenced by entire industries and branches, clusters turned out to be an effective tool facilitating and stimulating entrepreneurship progress, stimulating and maintaining competitiveness, and a mechanism for a competitive strategy. They became an approach for fostering innovation, co-operation and internationalisation among companies and regions to succeed in global competition.

Popularisation, as well as the vivid and growing presence of the cluster phenomenon over the last 20 years, has contributed to the induction of numerous researches focused on the issue of clusters and their development around the world with particular attention to European countries.

This book presents the current results of research conducted in the area of clusters and cluster policy development in Central European countries with a focus on Poland, the Czech Republic, the Slovak Republic and Austria. Austria is shown to be a country with an advanced economy, with a developed cluster policy and a significant proportion of established clusters. Other countries are part of a territory which experienced similar political and economic development over recent decades. Clusters in the Czech Republic, Poland and the Slovak Republic are largely at the stage of embryonic clusters, where the government is in the process of setting up the conditions for effective cluster support.

The aim of this book is to map and compare the conditions influencing cluster development in the selected countries from both the macro (policy, government) as well as the micro (cluster organisation) level, analyse good practices and propose a framework for cluster development in this region. In this context, the book serves as a guide for successful cluster development and a source of valuable thoughts aiming to broaden the horizons of knowledge in regard to the issue of clusters in the selected region.

The research presented within this book brings a new perspective on the cluster issue and extends the investigation which had been carried out within cluster-oriented projects implemented by the Department of Finance and Accounting and Centre for Applied Economic Research, at the Faculty of Management and Economics (FaME), Tomas Bata University in Zlin (TBU). The very first, leading project was entitled "Cluster Performance Measurement and Management" (project No. 402/06/1526) implemented under the support of the Grant Agency of the Czech Republic in the years 2006-2008 and conducted by Drahomíra Pavelková. Within it, the set of data depicting the status of cluster development in the Czech Republic and worldwide was collected in 2007, constituting the initial data for further cluster-oriented research at TBU. Data was collected using a questionnaire survey and structured interviews.

The outcomes of conducted research build on the findings introduced as part of the results of consecutive research projects implemented under the support of the Internal Grant Agency of the Faculty of Management and Economics at TBU, the International Visegrad Fund, Technology Agency of the Czech Republic and the Operational Programme for Education and Competitiveness. The investigation was conducted within the following projects:

- Internal Grant Agency of FaME TBU No. D/09/5: "The analysis of clusters' background based on Central Europe countries of Poland, Czech and Slovak Republics" (2009) conducted by Magdalena Bialic-Davendra;
- International Visegrad Fund No. 50910431: "Clusters performance measurement and management in Poland" (2009/2010) conducted by Magdalena Bialic-Davendra;
- Internal Grant Agency of FaME TBU No. DP 9/10: "The investigation of a new concept of World-Class Clusters in Central Europe" (2010) conducted by Magdalena Bialic-Davendra;

- International Visegrad Fund No. 51000656: "Clusters performance measurement and management" (2010/2011) conducted by Magdalena Bialic-Davendra;
- Internal Grant Agency of Tomas Bata University in Zlin No. IGA/61/FaME/10/A: "The Development and Evaluation of the Performance of Cluster Policies, of Clusters and their Members with the Usage of the Principles of Benchmarking" (2010-2012) implemented by a project team with the project leader Eva Vejmělková (Jirčíková);
- Technology Agency of the Czech Republic, programme Omega, grant No. TD010158: "Cluster policy of the Czech Republic and its regions for global competitiveness and sustainable growth" (2012-2013) implemented by a project team with the project leader Drahomíra Pavelková;
- Operational Programme Education for Competitiveness No. CZ.1.07/2.3.00/20.0147: "Human Resources Development in the field of Measurement and Management of Companies, Clusters and Regions Performance", conducted by Drahomíra Pavelková.

The evidence as depicted in this book are based on the findings presented within the doctoral thesis entitled "Identification of Factors Influencing the Cluster Development Process in the Czech Republic" defended by Eva Vejmělková (Jirčíková) in 2008 and "Proposal of a framework for a cluster development in selected Central European countries" defended by Magdalena Bialic-Davendra in 2011, both at Tomas Bata University in Zlin.

In order to obtain specific information regarding the present stage of development of the cluster concept and cluster-based policies in the selected countries, the survey with the use of tools such as telephone interviews (as the first contact) and questionnaires were conducted. The survey was directed at managers of a cluster/cluster initiative and representatives of institutions/agencies supporting cluster development. All the questionnaires were prepared in three language versions of English, Polish and Czech in order to enable easier availability and increase the efficiency of the conducted research. The questionnaire survey was supported through structured interviews with managers and/or representatives of the chosen clusters and representatives of the institutions supporting cluster development. Those, in turn, were not only the source of broadened knowledge regarding the current situation concerning the cluster concept and its development but also an inspiration for undertaking further steps and deepening research devoted to this issue. Data was collected in three waves from 2007 to 2011.

The results of both the critical research of the information sources based on professional publications, professional studies and case studies primarily devoted to the selected countries as well as practical research based on the survey and face-to-face interviews are presented in a total of five chapters. The sixth chapter constitutes the conclusions.

The first chapter is devoted to introducing the issue of clusters and their concept. For the purpose of the book, both the definition of a cluster and a cluster initiative are clarified. Furthermore, cluster characteristics and attributes are depicted. In addition, attention has been directed to the importance of cluster evolutionary growth i.e. the cluster life cycle and its activities following a certain path of cluster development.

In the second chapter, attention has been given to the dimension of policy based on clusters, where the term "a cluster policy" is defined. In addition, the development of cluster policy in the European Union over a span of time is described. The character of the cluster policy is also depicted with the example of selected European countries i.e. Poland, Czech Republic, Slovak Republic and Austria, with their comparison being conducted.

The importance of clustering for companies and other entities involved in the cluster is presented in the third chapter. As joint activities of cluster members are inseparably connected with the development of the cluster as an entire organisation and serve as a trigger for the realisation of its goals, the conducted research focused its attention on their specificity (also in connection with the country of origin of the clusters). Furthermore, examples of cluster best practices are depicted. On a basis of critical literature research and practical research (interviews and questionnaires) conducted within this book, (key) factors considered the most significant in creating an ideal environment for clustering were identified and the framework for cluster development in selected Central European countries was formulated in chapter four. Additionally, special attention has been devoted to the activity of benchmarking as an example of a tool for assessing the performance of clusters and their members.

As globalisation and internationalisation processes have a profound influence on the perception of competition and co-operation, thus on clustering itself, chapter five devotes its attention to new concepts of inter-cluster co-operation and World-Class clusters featuring their aspects.

Finally, the sixth chapter summarises the outcomes of the conducted research presented in this book.

Last but not least, we would like to use this occasion to thank everyone who supported our research, i.e. cluster managers and cluster representatives as well as representatives of institutions and agencies supporting clusters, for their dedicated time and co-operation.

In advance, we would like to thank you all for sharing your valuable comments, thoughts, experience and any further ideas connected with the published topic with us (at the following e-mail addresses: bialic.davendra @gmail.com or bialic@fame.utb.cz).

CHAPTER ONE

THE NATURE OF CLUSTERS

The phenomenon of clusters embraces Alfred Marshall's (1920) concept of industry districts dating back to the end of the 19th century. This British economist noted that particular industries tended to concentrate locally on small companies with similar or complementary profiles thus bringing substantial benefits from the externalities for the region and the industry itself (Bialic-Davendra and Pavelková 2010).

Other approaches such as Perroux's theory of growth poles, the theory of industrial location of Weber and Hoover, or the path dependence theory have also influenced the development of the cluster concept (Błasiak-Nowak 2007).

At present, the cluster phenomenon is growing in popularity, becoming prevalent in supranational and national policies and strategies for the economic development of countries. It is perceived as an important development tool by world organisations such as the Organisation for Economic Co-operation and Development (OECD), the United Nations Industrial Development Organisation (UNIDO), the World Bank and the European Union (EU), amongst others.

Defining a "cluster" does not seem to be a simple task. As its concept refers to a variety of business structures and reflects various purposes, numerous definitions have been formulated including a cluster initiative, industrial clusters, regional clusters, knowledge clusters, research-driven clusters, cross-border clusters, world-class clusters and many others (Anderson et. al. 2004, Blien and Maier 2008, European Commission FP7[1], Evers 2008, Sölvell, Lindqvist and Ketels 2003). Regardless of the definition, clusters are depicted as an effective tool stimulating the development of their members and the surroundings.

---

[1] http://cordis.europa.eu/fp7/capacities/regions-knowledge_en.html (accessed May 5, 2011).

Definition of a cluster concept (cluster vs. cluster initiative)

As literature has indicated, over the last 20 years numerous researchers have tried to define the term "cluster" using their own perspective (Porter 1990, Rosenfeld 1997, OECD 1999, Mytelka and Farinelli 2000, Cooke 2002, Enright 2003, Grycuk 2003, Sölvell et. al. 2003). It would appear that no standard definition has been agreed upon. Most definitions, however, underline the geographical proximity of the companies, their interactions, and the horizontal and vertical connections.

In 1990: 213, Michael Porter introduced his definition of a cluster as a:

> geographical concentration of interrelated companies, specialised suppliers, service providers, businesses operating in similar sectors as well as related institutions (like universities, normalization organisations and institutes as well as branch associations). In certain fields, these organisations collaborate and compete as well.

which undoubtedly made the greatest impact on the current understanding of the cluster concept. Since that time, the concept of clusters has begun to arouse interest amongst entrepreneurs and policy makers.

Later in 1998: 78, Porter extended his definition describing clusters as:

> geographic concentrations of interconnected companies and institutions in a particular field. Clusters encompass an array of linked industries and other entities important to competition. They include, for example suppliers of specialised inputs such as components, machinery, and services, and providers of specialised infrastructure. Clusters also often extend downstream to channels and customers and laterally to manufacturers of industries related by skills, technologies, or common inputs. Finally, many clusters include governmental and other institutions–such as universities, standards-setting agencies, think tanks, vocational training providers, and trade associations–that provide specialised training, education, information, research, and technical support.

A similar view of the cluster concept is depicted by the CLUNET Policy Guidelines Report[2] (:2) which characterises clusters as:

---

[2] PRO-INNO Europe. "CLUNET Policy Guidelines Report". http://observatoirepc.org/fileadmin/user_upload/Ressources/CLUNET_policy_Guidelines_080108.pdf (accessed December 8, 2010)

> flexible networks of small and large companies that complement each other, enhanced by research, development, qualification institutions and additional centres of competence that build competitiveness thanks to close supply linkages and co-operative relationships.

Additionally, they:

> occur due to proximity to markets, in the presence of specialised labour, inputs (natural resources, information) and equipment/service suppliers, and due to the availability of infrastructure[3].

The geographical range of clusters can include a group of neighbouring countries, one country, region, voivodeship, district, community or even a city (Wójcik-Augustyniak 2009).

Another cluster definition underlining the importance of networking linkages was the one introduced by the OECD in 1999, stating that:

> clusters can be interpreted as reduced-scale national innovation systems with their "focus on knowledge linkages and interdependencies between actors in networks of production". They are "often cross-sectoral (vertical and/or lateral) networks, made up of dissimilar and complementary firms specialising around a specific link or knowledge base in the value chain (OECD 1999:8).

Furthermore, UNIDO (2001: 9) defines clusters:

> as sectoral and geographical concentrations of enterprises that produce and sell a range of related or complementary products and, thus, face common challenges and opportunities.

Also, the European Commission (2005: 9), on the basis of Porter's definition, describes clusters as:

> groups of independent companies and associated institutions that are:
> - Collaborating and competing;
> - Geographically concentrated in one or several regions, even though the cluster may have global extensions;
> - Specialised in a particular field, linked by common technologies and skills;
> - Either science-based or traditional;

---

[3] OECD. "LEED-CEI-EBRD project on Industrial Districts and Local Clusters 2001-2002". http://www.oecd.org/document/6/0,3343,en_2649_34457_2438406_1_1_1_1,00.html (accessed February 27, 2011).

- Cluster can be either institutionalised or non-institutionalised.

Clusters specifics is also emphasised in further studies, for example, in the British report[4] from 2004: 4-5, as groups of inter-related industries where:

> the links between firms are both vertical, through buying and selling chains for example, and horizontal, through complementary products and services, the use of similar specialised inputs, technologies or institutions, and other linkages for example. Most of these linkages involve social relationships or networks that produce benefits for the firms involved.

Entities inter-related within a cluster "pursue the same development trajectory (for example target markets, technology)" (Szymoniuk 2005).

Also according to Stachowicz (2006: 28), "clusters are innovative systems in which the linkage between business and research-and-development sphere plays a crucial role".

As Gulda (2008: 8) states:

> It seems that clusters constitute one of the interesting solutions which can unite the efforts of a group of entrepreneurships, strengthen them through finding a common platform of agreement, common goals which exceed beyond competitive aims. They are a potential key aspect of regional value as they provide a unique combination of knowledge, skills and abilities significant for the competition site (Kačirková 2008: 451).

Following the Gorynia and Jankowska's (2008) definition, clusters can be characterised as an example of the mezzo-economic system placed between the national (macro-system) and the entrepreneurship (micro-system), where all the essential "ingredients" are joined in order to achieve competitive success.

Along with the definition of "a cluster", the term "cluster initiative" (CI) often appears as inseparably connected with cluster oriented research. According to the Cluster Initiative Greenbook[5] (:9), CI can be interpreted as "organised efforts to enhance the competitiveness of a cluster, involving private industry, public authorities and/or academic institutions". Most

---

[4] "A Practical Guide to Cluster Development. A Report to the Department of Trade and Industry and the English RDAs by Ecotec Research & Consulting", England's Regional Development Agencies, 2004,
http://www.dti.gov.uk/files/file14008.pdf (accessed May 20, 2010).

[5] Sölvell, Örjan, Lindqvist, Göran and Christian Ketels 2003, *The Cluster Initiative Greenbook*. Stockholm: Ivory Tower AB.

cluster initiatives emerge in the form of a project and are converted into a more formal structure within 12 to 36 months (Ketels, Lindqvist and Sőlvell 2007). CI's provide an initial platform for clusters to evolve.

Apart from these defined clusters/cluster initiatives, additional phenomena of cluster structures can also be distinguished as:

- Entrepreneurship environments, which are currently in the organising process, with no defined co-operation network among the participants;
- Platforms of co-operation with the nature of a prestigious business club;
- Projects or institutions, which are in a developing process at the present moment and are not yet clusters or were wrongly associated with a cluster organisation*.

These cluster structures were excluded, however, from the research presented within this book.

Since the clusters analysed within this book may in certain cases indicate the characteristics of a cluster initiative, it was consequently necessary to specify the term "cluster" used for the purpose of this publication.

> A "cluster" is defined as all forms characterised by an organised structure, concentrated around a dominant sector, operating both in the production, service or agricultural sectors, aware of their territorial identity, which carry out joint initiatives, and which also take into consideration those showing the characteristics of a cluster initiative.

Cluster characteristics

Each cluster is unique and has its own attributes. Nevertheless, certain common characteristics can be distinguished, which apply to clusters as a structure (organisation).

According to Porter (2008), clusters vary in size, breadth and state of development. Some consist primarily of small and medium-sized enterprises (SMEs), others involve both large and small firms; some are centred on research universities, while others have no important university connection. The differences in their nature reflect the differences in the structures of their industries. They differ in a number of dimensions. A number of them may be found in tight groupings concentrating several companies in a small geographical area, while others spread across regions with a large density of entities. Clusters can exist with numerous linkages, activity-rich, concentrating a number of related industries, whereas others are involved in one or only a few activities in a given industry. There are science-driven clusters with high innovative capacity which focus on technological activities and those in traditional sectors with a lower innovation ability; well-established and just emerging clusters (Enright 1998 and 2003, Ketels, Lindqvist and Sölvell 2008).

In The Cluster Policies Whitebook (Andersson et al. 2004: 1), the authors describe seven elements as typical cluster characteristics:

1) Geographical concentration;
2) The specialisation or common denominator of a cluster;
3) The cluster actors;
4) Cluster dynamics and linkages: competition and co-operation;
5) Critical mass;
6) The cluster life cycle;
7) Innovation.

They underline the geographical proximity of various actors (multiple organisations), which are linked within a cluster by co-opetition[6], with their focus on a common core business (activity).

Additionally, Gorynia and Jankowska (2008) highlight the significance of relative proximity of entities, usually high density-intensity of economic activity, the presence of numerous firms engaged in the same, similar or substitution activity, and the connection of the cluster formation process with the historical conditions of the area in which the cluster emerges. Consequently, a certain level of specialisation or common denominator of a cluster, which links entities within various forms of co-operation (for example within the value chain), appears as a further attribute.

A cluster by definition consists of particular actors linking the spheres of industry, education and government, so called Triple Helix

---

*According to CzechInvest (2005), a cluster organisation is a specific legal body established with the purpose to manage a cluster.

[6] Co-opetition is a neologism coined to describe co-operative competition; it is a conjunction of co-operation and competition (Brandenburger and Nalebuff 1996).

(Etzkowitz 2002). According to Sölvell, Lindqvist and Ketels (2003) five sets of actors can be differentiated. Figure 1-1 presents the cluster and its actors representing different spheres of a Triple Helix.

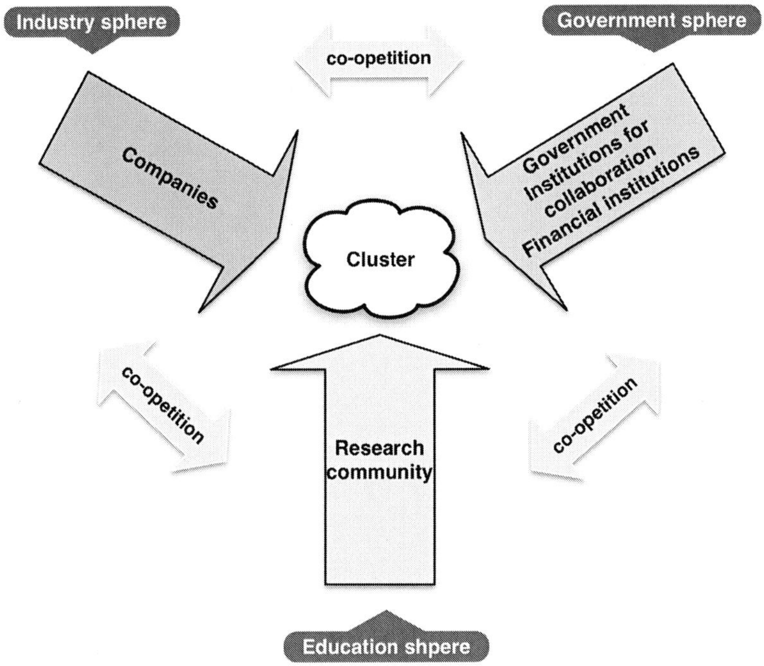

Fig. 1-1 The actors composing a cluster
Source: Own development based on Etzkowitz (2002: 4)
and Sölvell, Lindqvist and Ketels (2003: 18).

The entrepreneurship sector is represented by buyers and suppliers of goods and services, companies sharing common technologies, labour skills, etc. Government includes national, regional and local institutions, agencies and other units involved in industry and economic development policies. Financial institutions encompass banks, venture capitals, business angels, etc. The research community is represented by academic actors such as universities and colleges, research institutions, technology parks, etc., and Institutions for collaboration (IFCs) constitute public agencies, public-private organisations such as formal and informal networks (Sölvell et al. 2003, Sölvell 2009).

Those actors are linked by the bonds of co-opetition within a cluster organisation. The cluster's function is based on a balance between competition and co-operation, while the co-operation amongst companies is based on trust (Bialic 2008). The linkages amongst entities result from transactions between producers and traders, manufacturers and organisations delivering business-supporting services, research and development facilities, scientific centres, etc. Companies inter-linked within a cluster organisation may bring along externalities in the form of resources or services, which would not be available for individual entities.

The clusters' regime is depicted as a relationship among its actors based on the Triple Helix model implemented by Etzkowitz (2002) i.e. linkages between industry, research (academia) institutions and government agencies, with its further expansion into the Quadratic Helix including additionally, the finance sector. The new approach of the Triple Helix (Figure 1-2) puts things into a different perspective highlighting the importance of the relationship between creativity, entrepreneurship and eco-systems (Quinn 2009).

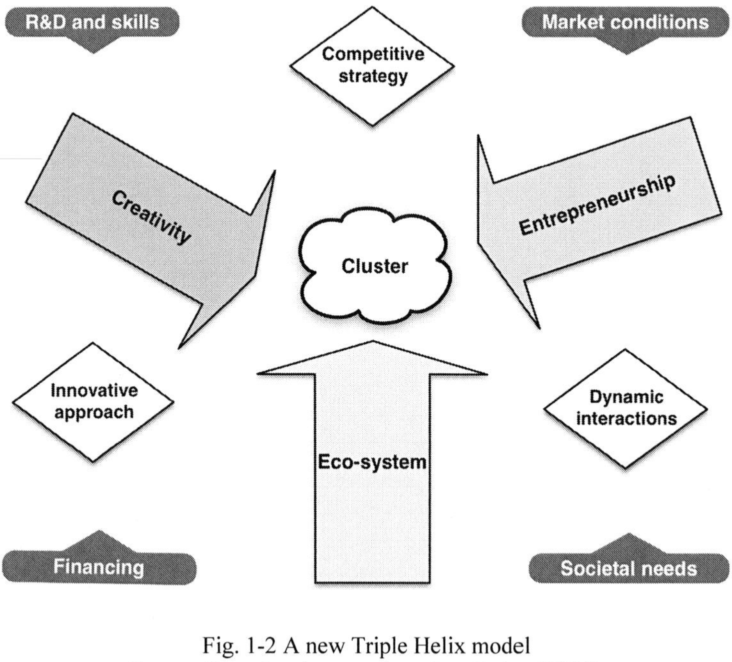

Fig. 1-2 A new Triple Helix model
Source: Own development based on Quinn (2009).

Increased attention is being drawn to creativity and the capacity of the correct understanding of the clustering idea (know-why concept) as the way of achieving a competitive advantage above others. Actors, particularly those representing the industry sphere, have to be flexible in terms of social conditions, economic activity, resource usage, etc., and offer unique features, which diversify them from others and result in being more attractive to the present market. New values begin to play an increasing role and clusters as separate entities are no longer taken into consideration but instead entire eco-systems. The new approach of inter-cluster co-operation or the process of "clustering the clusters" (co-operation among whole cluster organisations) appears as a modern tool for a competitive strategy, where the level of co-operation is being expanded from trans-regional to trans-national, thus stimulating the innovation, growth and competitiveness on a wider scale.

The authors of the "Whitebook" also pinpoint the fact that clusters need to be innovative and engage a number of actors (a critical mass) in order to be dynamic and multiply interactions. It is important that multifaceted relationships occur amongst cluster members and participants. These relationships have to exist respectively between a large number of entities creating the cluster organisation, where the so-called "critical mass" must be exceeded. The aggregation of interrelated entities, which have not reached the critical mass, nevertheless constitutes a potential cluster, which may or not be transformed into a cluster structure (Wieczorek 2008). As Sonderegger and Täube (2010: 389) clearly state:

> Once the cluster reaches the critical mass, it enters into a phase of (rapid) expansion and sparks the exploration of new products, processes and business models. Firms systematically get to know about innovations and strategic changes that have been developed or are executed by their competitors, suppliers and customers (Bathelt 2005: 109).

New entities appear to be attracted by the growing cluster. The accomplishment of the critical mass enables the obtaining of synergy effects for cluster participants including the diffusion of knowledge, innovations and technology, concentration of resources leading to an increase in their performance, attracting new resources and entities into a certain location, etc. (Pilarska 2009).

Clusters stimulate and revitalise co-operation in the business environment. Entrepreneurships co-operating within such a network can gain multiple benefits derived from various activities. As the competitiveness of entrepreneurships strongly depends on their level of innovativeness, companies feel the need to provide sophisticated methods and tools that enhance the innovation technology development in the company. As Gorynia and Jankowska (2008) aptly note, a similar level of technical advancement has been observed as an attribute of a cluster. Clusters stimulate competitive pressure even among indirectly competing or non-competing participants, what has a motivating influence on entities, which are thereby more likely to co-locate R&D, co-operate on production, co-monitor and rapidly perceive customer needs, carry out common marketing activities, etc. (Porter 2008).

Another characteristic element is the recognisable life cycle, which is described in detail in the next section.

A further important feature is the way in which clusters emerge. Gorynia and Jankowska (2008), following previous research (Saxenian 1994, Pinch and Hendry 1999, Feldman 2001, Zander 2003) distinguish two leading approaches:

1) Bottom-up approach–where a cluster emerges due to the activity and/or own initiative of groups of entrepreneurships;
2) Top-down approach–where a cluster emerges as a result of the activities of economic policy.

A third approach may appear where a cluster emerges due to a combination of both specified approaches and/or due to the historical conditions (Hendry, Brown and Defillippi 2000).

Cluster establishment may be stimulated by the public authorities or it can be a result of interest and enthusiasm on the part of the companies themselves. However, the creation of clusters should constitute the market process, not an administrative decision. The role of public authorities should be limited to indirect support in the form of, for example, support of the infrastructure, the development of an appropriate focus within the educational system, the creation of business incubators, technological parks, the rapid adoption of regional strategies and spatial plans or promoting a cluster brand. They should play the role of a catalyst, where their support should be limited along with the development of a cluster.

All the distinguished characteristics are common for clusters, nevertheless, their presence is not compulsory within cluster structure.

In addition, there are numerous typologies of clusters, which emphasise various dimensions of their variability (Table 1-1).

Table 1-1 Typology of clusters according to the criterion.
Source: Own development on the basis of Skawińska and Zalewski (2009), p.180; Gorynia and Jankowska (2008), Rugman and Verbeke (2003), Brodzicki and Szultka (2002), Hegedüs (2008).

Criterion	Type
Company size	Large enterprises, small and medium enterprises, mixed (hub-and-spoke)
Life cycle phase	Embryonic, established, mature, declined
Period of development	Learning co-operation, maturity, decline period, globalisation (merger, take over)
Type of activity	Industrial, agricultural, service, building, IC
Cluster dynamics (level of flexibility and aggregation of capabilities)	Static, dynamic*
Duration of existence	Newly established, secondary (revitalised)
Geographical coverage	Concentrated, distributed
Degree (depth) of co-operation	Shallow, deep
Number of linkages of sectors (breadth)	Narrow, wide
Technology importance	Local innovative clusters (high, medium, low) and traditional
Level of technological advancement of the activities undertaken	High level, low level
Innovative capacity	Highly innovative, weakly innovative
Cluster organisation (structure)	Stem with company co-ordinator, stem with company-leader, lack of co-ordinator and/or leader
Orientation and scope	Domestic and foreign, symmetrical and asymmetrical

Main factor	Resources (raw materials, conditions), intellectual (quality of human capital, quality of environment), with foreign capital
Growth potential/market position	With increasing potential and strong market position, with stabilised potential and strong or average market position, with decreasing potential and weak market position
Dominant mechanism of coordination and regulation of member behaviour	Transactions ad-hoc, short-term coalition, long-term relations, hierarchy
Type of a cluster network specific for the region or country	American, Italian, Danish, Dutch

*A cluster with the dynamic capabilities and flexibility to innovate, incorporate new technologies, tackle new markets, etc.; with ability to easily adapt to the changing (business) environment (Sölvell, Lindqvist and Ketels 2003, O'Callaghan 2007, Szanyi et al. 2007).

Companies within the cluster carry out business in a common field with their production being often mutually complementary. In unison, they are able to respond faster to market changes and meet more demanding requirements of their customers. Additionally, sharing marketing activities, better and faster access to information and new technologies provides better opportunities for the rapid development of the companies. Furthermore, mutual co-operation inside the cluster brings cost reductions and benefits from economies of scale. In addition to this, a cluster as a group of companies has a stronger negotiating power in sales and purchasing and a greater potential to receive support from governments and/or for example from EU funds (Staszewska 2007 and 2009). Clusters through increasing competitiveness and performance of companies, innovation incentive (for example effective R&D, support of spin-offs, attracting foreign investments), serve to stimulate the economic development of national economies.

Cluster life cycle

The relevant literature describes markets and companies (even products and technologies) going through their developing, growing and declining stages, where Klepper's (1996) industry life cycle model (see also Audretsch and Feldman 1996, and Klepper 1997) or the Utterback and Abernathy (1975) technology life cycle model (see also Dalum, Pederson and Villumsen 2005) can be distinguished.

Clusters also evolve over time, where various cluster activities can be associated with different phases of their development. According to Menzel and Fornahl (2009: 211-212):

> …geographical cluster as such is only a theoretical concept that describes a particular constellation of different elements (for example companies, organisations, and networks) and processes (for example interaction and monitoring), therefore, the movement of the cluster through the life cycle is not performed by the cluster, which is only a concept, but is the result of the activities and the evolution of its elements.

Cluster activities usually follow a certain path and have a place throughout a cluster's life cycle (Pavelková et al. 2009).

Different perspectives on the life cycle of clusters and defined stages of their development are described in the relevant literature, often contrasting the importance of a cluster's evolutionary growth with its disregard (Lorenzen 2005, Boschma and Frenken 2006).

According to the authors of "A Practical Guide to Cluster Development" (2004), the following four stages of development can be distinguished within a recognisable cluster life cycle:

- **Embryonic clusters,**
- **Established clusters,**
- **Mature clusters,**
- **Declining clusters.**

The Embryonic (first) **phase** of a cluster life cycle characterises the beginning stage of development of co-operation linkages among cluster actors. As investments in the cluster and in its critical mass take place during this phase, the role of a supporting policy (the European Union and national funding for cluster initiatives) is crucial. Further growth is synonymous with **established clusters**, whose products and services are fairly established and where a gradual increase in the participation of new members can be observed. This stage is the prelude to the internationalisation of the cluster concept as more and more clusters venture into the international market. The major catalyst for this is superior internal dynamics and the attraction of resources from the outside-circulation of ideas, skills and resources. The presence of a mixture of advantageous diamond factors (Porter 1990) is needed in this stage. The pinnacle stage of the cluster life cycle is **maturity**. These clusters have reached the maximum development and exploitation phase, and have stagnated. The standardisation of the processes and products is a typical feature of this phase. Competitive pressure is an ever increasing presence. The **declining stage** is concerned with clusters which have reached their peak and lose their potential and significance (they are failing or declining; products and the services of the cluster are replaced by other cheaper substitutes), or with the possibility to innovate and re-start the cycle again. The cluster closes in its own environment, which was earlier constituted of its strength, for example, a qualified work force, unique knowledge, strong connections with other entities. It can fall into the trap of rigid specialisation. The cluster can experience either a lock-in effect (Grabher 1993-used the metaphor of "the weakness of strong ties" to describe this effect; see also Hassink 2005) losing the ability to adjust to a changing environment or a renaissance involving entering new growth phases by integrating and applying new technologies and knowledge (Bergman 2008, Górzyński 2006, Martin and Sunley 2006, Menzel and Fornahl 2009, Staszewska 2009).

Sölvell (2009) discusses the fact that clusters experience all these things in common with every social system: **birth, growth, decline** and **death**.

Each cluster has its unique life cycle and evolution (see also Saxenian 1994, Tel Wal and Boschma 2011). Different perspectives on the life cycle have been depicted by Pouder and St. John (1996), Tichy (1998), Swann (1998) and Brenner (2004).

Wolter (2003) defines the life cycle for clusters in a more generic sense, encompassing all forms of regional concentration of economic activity which develops the life-cycle theory of agglomerations with the **set-up**, **growth**, **change** and **adaptation** stages of development.

The four stages of cluster development which have been defined by Van Klink and De Langen (2001) outline certain related characteristics such as the character of the value chain, strategic relations, cluster dynamics, the co-operative domain, the determinant for success and the government's role in enhancing clustering. The **development phase** creates a link amongst actors and sets up a value chain. This leads to strategic relationships among companies within a cluster, as they need one another's products and services. The development of R&D activities, co-operative routines and standardisation takes place with a focus on new product development and the improvement of production methods. This phase is followed by the **expansion phase** which is linked to growth through emerging markets, R&D, education and marketing with a focus on the production process and the expansion of the product range. Sharing of the infrastructure (physical, information or knowledge infrastructure) is becoming more common. The **maturation phase** is where the cluster has an established product or set of products and a stable plan for production and sales. In this stage, the activities include education, research and development and marketing. In the **transition phase** either a decline of a cluster or the development of new co-operation can be observed. Clusters focus on R&D, education and/or new co-operative routines.

Žižka (2006), following the division depicted in the "Whitebook" (Andersson et al. 2004), describes five different stages of the cluster life cycle, these being:

1. **Agglomeration** – where a number of companies and additional actors exist, their co-operation can be spontaneous or the result of a consensus of co-operation.
2. **Emerging cluster** – the actors begin to co-operate around a core activity; the creation of formal or informal network structures is observed; companies begin to implement shared projects; a cluster embryo is being created.
3. **Developing cluster** – new connections develop between all the actors, new actors in the same or related activities emerge or are attracted to the regions.
4. **Mature cluster** – a certain critical mass of actors is reached; the relationships outside of the cluster, to other clusters are being developed; the spectrum of shared activities is widening; new companies-spin-offs and strategic alliances are created; the cluster actors share their vision, the transfer of knowledge and co-operation among the private and public sectors can be distinguished.
5. **Transformation** – the phase of change in markets, technologies and processes; the stage in which clusters either innovate and adapt to changes or decline.

A contrasting approach is given by Maskell and Kebir (2005) who describe the cluster life cycle phases using a three-stage development: **existence, extension** and **exhaustion**.

Bergman (2008), in turn, similarly lists the phases of existence, expansion, exhaustion and extinction. In his view, not all clusters experience the life cycle in full. This view is further reinforced by Sonderegger and Täube (2010) who illustrate the process of the cluster life cycle in five stages as follows:

- **Pre-cluster foundations** – related to historical and institutional foundations;
- **Emergence** – where historical factors play a major role in cluster emergence, the formation of clusters can be triggered through local demand for certain products, technologies, etc.;
- **Exploratory growth** – when the cluster reaches a critical mass; rapid expansion of new products, processes and business models as well as fast innovation development taking place;
- **Exploitative growth** – when a cluster technology matures, incentives for innovation are reduced and a stable buyer-supplier relationship limit the extent of new knowledge inputs; the development of local networks and their strength is observed;
- **Exhaustion** – congestion occurs, which can appear as an excessive extension of the infrastructure, the exhaustion of labour, high entry barriers as well as the technological lock-in effect; external stimulus influencing changes are desirable; either a renaissance and the beginning of a new life cycle process, or a lock-in effect takes place based on the ability to adapt to changes.

Yet another view is presented by Malakauskaite and Navickas (2010 and 2011) who when discussing the connections between the level of clusterisation, the competitive advantage and the cluster life cycle, also distinguish between several phases of the cluster life cycle such as: **establishment**, **development** including slow growth (development) of a cluster and fast growth (development) of a cluster, **maturity, maturity transforming into the decline of a cluster, decline and transformation.**

Further view on clusters life cycles and cluster activities associated with different phases of their development in the four countries is being discussed in Chapter three.

CHAPTER TWO

CONDITIONS FOR THE EMERGENCE AND DEVELOPMENT OF CLUSTERS

The emergence and development of clusters has been influenced not only by the micro conditions shaped by the cluster itself, but also by the overall macro environment. The macro factors, starting from the available resources, through the historical and cultural background, up to the existence of economic and legal framework conditions in a particular region or country, constitute the basic fundamentals for clustering. Government efforts shaping such a cluster-based policy, which would stimulate co-operation, strengthen competitiveness and accelerate the clustering processes, appear as one of the extremely important aspects.

Cluster policy

As clusters join numerous diverse entities from different spheres i.e. industry, education, science, government, etc. into a consistent structure, it becomes necessary to take into account all the measures affecting their existence when defining a **cluster policy**.

According to Pavelkova et al. (2009), the core of a cluster policy lies in a series of different activities, strategies, programmes and procedures, etc., which pose a focal point for increasing the socio-economic benefits as a result of the existence and development of clusters. These steps are consistent with a certain plan and designated budget and are fulfilled during a defined time frame (usually several years). A cluster policy can be explicitly focused on clusters or it can indirectly affect clusters within a strategy focused on boosting competitiveness or general economic development (Kačírková 2008).

The European Commission (2005: 10) faithfully reflects the essence of a cluster policy which:

> is not an isolated, independent and well-defined discipline. It embraces all policies that affect the development of clusters, taking into account the

> synergies and interchanges between these policies. Many policies labelled under different headings (regional policy, industrial policy, innovation policy, etc.) are in fact cluster policies.

Cluster policy may combine a wide range of policy areas (Szanyi et al. 2007), starting with policies focused on innovation and technology, through regional economic development policy, entrepreneurship/SME policy, as well as many others such as competition policy, science, education and training policy, labour market policy, etc. They all create a sort of overall cluster policy mix.

The literature therefore often depicts the issue of a cluster-based policy. These types of policies:

> acknowledge the important role of public policy in shaping many dimensions that are important for company success and they stress the crucial role of rivalry between companies for innovation and for growth (Ketels and Memedovic 2008: 377).

Following these thoughts, Pilarska (2010) points out that a cluster-based policy includes all government efforts, both undertaken by the government itself or in co-operation with other entities directly targeted at clusters and their competitiveness (for example the influence on cluster management, financial support). In addition, the importance of government steps indirectly focused on clusters which can have a meaningful influence on their growth, i.e. connected with promotion, monitoring, reporting, etc., has to be highlighted[1]. As Ketels (2009) aptly notices, this concept has to be taken into account in a broader spectrum involving cross-cluster policy and its impact on clusters as well.

Cluster policy is also characterised according to its range for those ones which are generally affecting business conditions or strictly focused on clusters themselves, i.e. cluster-specific (Pilarska, 2010). Specific cluster policies appear to most often support those programmes and initiatives focused on upgrading human resources, cluster expansion, development of business and its environment, as well as on R&D and innovation[2].

---

[1] "Innovation Clusters in Europe: A statistical analysis and overview of current policy support". DG Enterprise and Industry Report, http://www.proinno-europe.eu/admin/uploaded_documents/innovation_clusters_in_europe.pdf (accessed February 27, 2011).

[2] Ibid.

According to the "Whitebook", different categories of cluster policy can be distinguished such as broker policies, demand side policies, training policies, measures for special promotion of international links and framework policies, which are complementary to specific cluster steps and which assist in accelerating the clustering processes (Andersson et al. 2004).

Andersson et al. (2004: 46) highlight that "the adoption of cluster policy may serve as a trigger for government and public authorities to alter out-dated governance mechanisms", and it should provide guidance to policy makers. In addition, it may also "help addressing institutional rigidities, for example, in universities and research institutes, and to increase the mobility of workers" (50). Nonetheless, when launching a cluster policy it is necessary to implement such mechanisms and measures, which should be effective and productive and lead to the achievement of added value.

According to the Expert Group, "cluster policy is developed by national governments in co-operation with regional or local governments" (European Commission 2005: 11). Due to the diversity of national conditions, the approach to cluster policy may differ. Nonetheless, common goals and characteristics can be pinpointed, such as the role of public and semi-public organisations as catalysts or mediators in stimulating clustering processes among companies. With their focal point on supporting the flow of information and knowledge, macro-level foundations and infrastructure as well as often ensuring financial assistance, these authorities attempt to assure an adequate environment for clusters/cluster initiative development.

Financial support and its influence on cluster development

A cluster as well as every other organisation, needs certain financial sources in order to run its activities. A sufficient budget is necessary in order to finance various projects (connected with research, promotion, education, etc.) and human resources (management, other cluster employees) without which the achievement of cluster objectives would not be possible. Nevertheless, additional financial sources coming from public or/and private funding may play a significant role particularly for new emerging clusters or cluster initiatives (they might be used to encourage companies to collaborate, for example, on a new product or process) and as a stimulating role for further faster growth of those clusters which are already mature (with developed links inside and outside the cluster).

According to the Cluster Management Guide[3], three types of financing can be distinguished:

- open-ended public funding (permanent subsidisation);
- temporary public funding (limited in time);
- public/private funding (public money in combination with services against payment).

The latest type of financing places an emphasis on changes in funding over time. This is the emerging stage of cluster development (early stage of cluster building) where public funding plays a leading role and where after a certain period of time (when further cluster growth is being observed) decreases come about in industry aid and cluster member funding. Often, co-funding clusters/cluster initiatives by authorities appear for a period of 18 months with the possibility of an extension for a further 18 months. Nevertheless, after three years' time, clusters should become self-sustainable, well funded and with no need for seeking separate funding.

For clusters to be considered mature public funding is not necessary, however, additional financial sources have a stimulating influence on their activities and bring better opportunities for further progress and strengthening of their competitiveness.

The study "A Practical Guide to Cluster Development"[4] underlines the importance of access to finance as it "contributes to the successful development of clusters through supporting the growth and expansion of cluster-related activities". Among the potential sources of financing public and private R&D funding, specialist resources (financial services), venture capital, business angels, and investor networks can be distinguished.

The European Union plays a major role in financial support of clusters and cluster initiatives as it creates the basis for cluster policy development in individual countries. Cluster policy on the EU level offers numerous programmes and initiatives with the possibility of obtaining funding.

---

[3] "Cluster Management Guide – Guidelines for the Development and Management of Cluster Initiatives", CLOE – Clusters Linked Over Europe, 2006, http://www.clusterforum.org/ (accessed May 31, 2010).

[4] "A Practical Guide to Cluster Development. A Report to the Department of Trade and Industry and the English RDAs by Ecotec Research & Consulting", England's Regional Development Agencies, 2004, http://www.dti.gov.uk/files/file14008.pdf (accessed May 20, 2010).

Cluster policy in the European Union

The introduction and popularisation of the cluster concept by Professor Michael Porter has brought about a vivid response in the development and adaptation of a cluster policy in Europe since 1990.

Although numerous cluster-based policies on national and/or regional levels specific for various countries have been developed, the cluster policy at the European level aims at uniting efforts for the strengthening and supporting of formation of clusters, networking and co-operation processes.

With the introduction of the Lisbon Strategy as a new development plan for the years 2000-2010, the European Union set a target to become:

> the most competitive and dynamic knowledge-based economy in the world capable of sustainable economic growth with more and better jobs and greater social cohesion[5].

The strategy pointed out innovations as the main engine for economic change where later development of a cluster concept would be meaningful. With the renewal of the Lisbon Strategy in the year 2005, the European Commission announced several strategic documents devoted to strengthening competitiveness and supporting clusters growth.

The presented "Integrated Guidelines for Growth and Jobs (2005-2008)"[6] highlighted the significance of common European initiatives stimulating the development of technology and co-operation between the private and public sectors, thus resulting in supporting establishment and development of regional and local clusters. A further focus on innovation activities and clusters was also introduced with the "Competitiveness and Innovation Framework Programme (2007-2013)" (CIP)[7]. Along with this, the Seventh Framework Programme (FP7) supporting the realisation of the EU goals on growth, competitiveness and employment through research-related EU initiatives has been implemented. Nonetheless, the main pillar

---

[5] European Union Parliament. Lisbon European Council 23 and 24 March Presidency Conclusion. http://www.europarl.europa.eu/summits/lis1_en.htm (accessed December 7, 2010).

[6] Integrated Guidelines for Growth and Jobs (2005-2008). Commission of EC. Brussels, 12.4.2005. COM(2005) 141 final.

[7] Decision No.1639/2006/EC of the European Parliament and of the Council of 24 October 2006 establishing a Competitiveness and Innovation Framework Programme (2007-2013). http://europa.eu/legislation_summaries/energy/european_energy_policy/n26104_en.htm (accessed February 27, 2011).

for the implementation of the European cohesion policy for the period of 2007-2013 is the "Strategic guidelines on cohesion"[8] which were constituted as the basis for the National Strategic Reference Frameworks (NSRFs) in individual European countries. These, in turn, have defined their own priorities for the use of the EU Structural Funds as well as their own operational programmes with possibilities of obtaining financial support for relevant projects implemented during the programming period. As stated in the "European Cluster Memorandum"[9], the CIP, the Structural Funds, and FP7 "include a wide range of activities open to clusters, innovation and competitiveness".

While implementing a cluster policy, the European Union started numerous programmes, projects and initiatives dedicated to co-operation and cluster themes and complementary to one another; Figure 2-1 depicts their areas of focus.

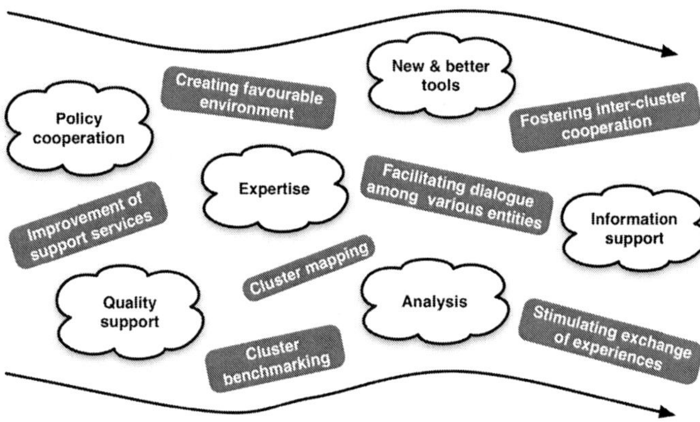

Fig. 2-1 The focus areas of the European Union projects and initiatives devoted to co-operation and clusters
Source: Own development

---

[8] Council decision of 6 October 2006 on Community strategic guidelines on cohesion (2006/702/EC). Official Journal of the European Union. L 291/11, 21.10.2006, http://ec.europa.eu/regional_policy/sources/docoffic/2007/osc/l_29120061021en00110032.pdf (accessed February 27, 2011).

[9] "European Cluster Memorandum", http://www.proinno-europe.eu/NWEV/uploaded_documents/European_Cluster_Memorandum.pdf (accessed December 7, 2010).

In 2006, the European Cluster Alliance (ECA) initiative was implemented with an aim to:

> maintain a permanent policy dialogue at EU level among national and regional public authorities responsible for developing cluster policies and managing cluster programmes in their countries.[10]

and simply to foster cluster co-operation. The ECA was founded by partners involved in the four cluster policy projects:

1) BSR InnoNet - the Baltic Sea Region Innovation Network – aiming to "create links between innovation policy makers, implementing agencies and analysts in the Baltic Sea Region", and to create an environment for building strong industrial clusters and innovation poles.
2) CEE-Cluster Network - Central and Eastern European Cluster and Network Area - linking "eleven partner regions/countries whose innovation policies focus on cluster and network policy" and shaping common policy.
3) CLUNET - Cluster Network - aiming to "share experiences and implement concrete pilot projects related to cluster innovation and development policies" through uniting 15 regions and 64 top clusters already supported by the partners.
4) INNET - Networking of national/regional funding and innovation organisations for the involvement of SMEs in technology-based innovation clusters in Europe - with a goal to "promote interaction and co-operation between regional, national and European funding schemes, and to facilitate cross-border co-operation for SMEs in technology clusters".

These projects are also known as INNO-Nets founded under the PRO INNO Europe initiative.

Another initiative is the European Cluster Observatory (further "the Observatory") launched in 2007. Its goal is to provide "information and analysis of clusters and cluster policy in Europe" as well as to improve statistical information on clusters (cluster mapping). It provides a variety of data regarding clusters in Europe, such as data and analysis on clusters and competitiveness, a library with reports dedicated to policy, regional clusters, cluster cases, good practices in cluster organisations, etc.

[10] http://www.proinno-europe.eu/eca (accessed May 5, 2011).

The Observatory also facilitates co-operation amongst cluster organisations and their members; it helps to find potential co-operation partners[11].

In 2009, the European Cluster Excellence Initiative[12] (ECEI) began its activity aimed at improving cluster policy within Europe and efficiency in cluster management (helping cluster managers achieve a high level of excellence). Within the ECEI framework, the European Cluster Managers' Club - Professional Association for Cluster Managers, was established with a goal of improving networking and cluster management skills. Another entity associated with the ECEI is the Europa InterCluster with the objective of promoting the exchange of experiences between European clusters.

The European Cluster Policy Group[13] (ECPG), also established in 2009, acts as an advisory body aiming to "explore how to better assist Member States in supporting the emergence of more world-class clusters in the EU". A group of 20 people called "the group of experts" focused their activity on analysing policies and financial instruments supporting clusters, identifying challenges for cluster policies and international trends for cluster development.

Another tool, the European Innovation Platform for Clusters (Cluster-IP, Cluster Partnerships), focuses on developing and testing improved cluster support services and integrating innovative SMEs into clusters. It aims to foster trans-national co-operation between sectoral cluster initiatives.

Another example of EU initiatives supporting clusters is the Clusters Linked Over Europe (CLOE), which is an informal alliance of clusters all across Europe. It was originally started as an INTERREG IIIC project in 2004. The clusters are linked within the alliance with the goal of exchanging information and working closely together in order to strengthen industry in a particular area. CLOE is also:

> fostering direct co-operation between cluster SMEs operating in the same industry, enhancing their understanding of their sector and helping them to identify future business partners[14].

---

[11] http://www.clusterobservatory.eu/index.html#!view=aboutobservatory;url=/about-observatory/ (accessed May 5, 2011).

[12] http://www.cluster-excellence.eu/3552.html (accessed May 6, 2011).

[13] http://ec.europa.eu/enterprise/newsroom/cf/itemlongdetail.cfm?item_id=2834 (accessed May 6, 2011).

[14] http://www.clusterforum.org/en/about_cloe.html (accessed May 5, 2011).

Alongside various EU initiatives, new documents shaping cluster policy development have also been introduced.

The European Cluster Memorandum (further "Memorandum") launched in 2008, has emphasised the need for further strengthening of co-operation and clusters. This brought a response in the Visegrad Group (further "V4") - countries of the Central European region i.e. the Czech Republic, Hungary, Poland, and the Slovak Republic, which in full harmony with the Memorandum and European Commission Communicate[15] have reached a "Memorandum of Understanding"[16] committed to supporting co-operation between clusters at national and international levels.

In March 2010, the European Commission introduced a proposal for a new strategy "Europe 2020: A strategy for smart, sustainable and inclusive growth"[17] including promotion of clusters within one of its seven flagship initiatives entitled "An industrial policy for the globalisation era".

The policy for clusters or cluster policies in the European countries can be classified as implemented in the EU macro-regions (for example Baltic Sea Region, Danube Region), national, regional, city-regions or local levels. All steps undertaken within these different levels of cluster policy should be complementary, should support and strengthen one another.

Despite the fact that clusters are "predominantly a regional or national phenomenon, the European level can contribute in a number of ways to their success" (Figure 2-2), through[18]:

1) Complementing regional and national cluster policies by removing barriers to trade, investment and migration, and thus deepening open competition across Europe;
2) Motivating and strengthening these policies, by developing and promoting a strategic approach to cluster policy in Europe;

---

[15] European Commission. Commission Communication "*Towards world-class clusters in the European Union: Implementing the broad-based innovation strategy*". COM (2008) 352 final.

[16] "Memorandum of Understanding on Co-operation in the field of the Cluster Development of the V4 Countries", Budapest, 26 November 2009. http://www.arr.cz/userfiles/file/Klastry/MOU_on_V4_Cluster_Co-operation.pdf (accessed December 7, 2010).

[17] European Commission. *EUROPE 2020. A European strategy for smart, sustainable and inclusive growth.* http://ec.europa.eu/eu2020/index_en.htm (accessed December 7, 2010).

[18] "Innovation Clusters in Europe: A statistical analysis and overview of current policy support". DG Enterprise and Industry Report, http://www.proinno-europe.eu/admin/uploaded_documents/innovation_clusters_in_europe.pdf (accessed February 27, 2011).

3) Supporting the creation of regional and national clusters by strengthening the knowledge base in Europe;
4) and stimulating the development of cross-border clusters, transnational co-operation between clusters.

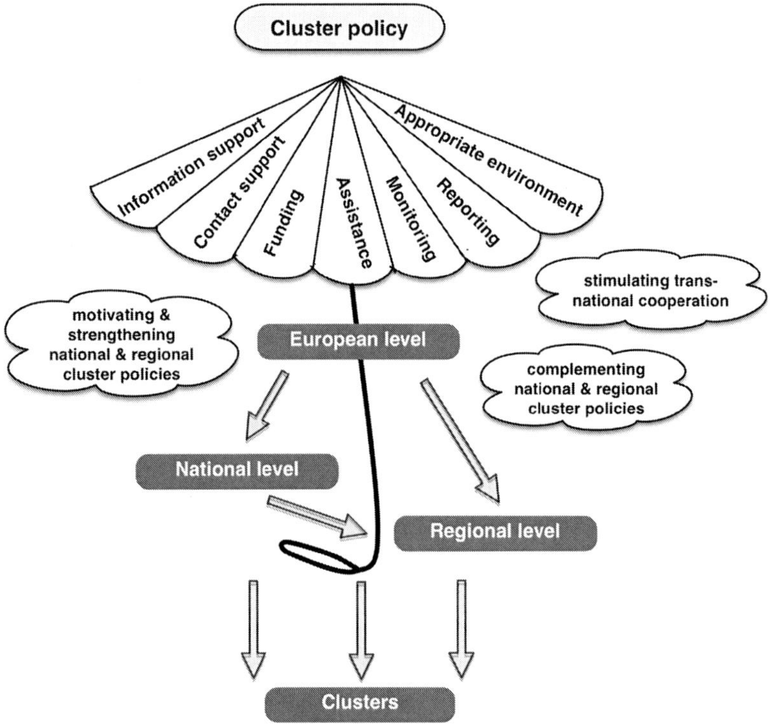

Fig. 2-2 Different levels of cluster policies
Source: Own development based on DG Enterprise and Industry Report. Innovation Clusters in Europe: A statistical analysis and overview of current policy support. p.21

The existence of various types of organisations supporting clusters contributes to ensuring an appropriate environment for the development of co-operation among companies and supporting entities i.e. universities, technological parks, R&D centres, financing institutions, etc. Various aid schemes depend on a level of engagement from the entities on the local or regional level as well as on national or even transnational institutions, which provide a different impact on networking and clusters development. In addition, various methods of support can be used such as:

> hands-on methods, like providing information, contacts, assistance, advice or direct funding to hands-off methods, like lobbying, marketing, monitoring and reporting[19].

Cluster policy in the example of Poland, the Czech Republic, the Slovak Republic and Austria

Schemes of cluster policies in the EU countries differ according to the level of engagement of governments (on national and regional levels). Most clusters arise or grow thanks to initiatives at the regional level. There are also examples, however, where national or even international institutions play an active role in cluster development (Mikołajczak, Kruszewska and Fila 2009).

As part of the Europe INNOVA, a Cluster Mapping Project was conducted, gathering information about cluster policies, institutions and programmes in 31 European countries[20]. The conducted research stated that all the countries have programmes focused on clusters either on a national or regional level, although the time periods for adopting a cluster policy differ and can be divided as follows: 1990-1994, 1995-1999, 2000-2004 and after 2005.

Cluster policy in Poland

An official cluster policy in Poland can be defined as non-existent and clusters are treated as one of the many instruments influencing economic development. Nevertheless, cluster development and support takes place within other policies (for example innovation policy, regional development policy, policies focused on SMEs development). All the activities undertaken towards clusters are part of the innovation development on a national and regional level.

---

19 "Innovation Clusters in Europe: A statistical analysis and overview of current policy support". DG Enterprise and Industry Report, http://www.proinno-europe.eu/admin/uploaded_documents/innovation_clusters_in_europe.pdf (accessed February 27, 2011).

20 "Cluster policy in Europe. A brief summary of cluster policies in 31 European countries", Europe Innova Cluster Mapping Project, 2008, http://www.clusterobservatory.eu/upload/Synthesis_report_cluster_mapping.pdf (accessed January 20, 2009).

As Gulda (2008: 8) aptly points out:

> Cluster policy does not constitute a separate branch of policy in Poland and there is no government document which would directly speak about it. Cluster policy is treated as part of innovation policy. The most common reference to cluster issues can be found in strategic documents connected with innovations. He also underlines that "clusters do not pose a separate subject of national policy but are an extremely important instrument which allows for combining and implementing other types of policy: innovative, connected with regional development and with the research and development sphere".

Szultka (2008: 13) also accurately states that:

> There is a lack of a comprehensive national policy, which would support clusters as a whole in such a way as to co-ordinate policy instruments from various areas towards the clusters' angle. Clusters in Poland are instead treated as one of many instruments.

The reason may lie in the fact of its comparatively young history and limited experience concerning clusters and cluster initiatives.

The first programme in Poland that had an influence on networking within Small and Medium Enterprises was the SME Clustering and Networking Programme[21] launched in 1995. Cluster-based policy was introduced for the first time, however, in 2002, when the Gdańsk Institute for Market Economics (IBnGR) began its research focused on analysing both opportunities (potential) and barriers for cluster structure development in the Polish economy, and on identifying the aggregations in economic activity. It analysed 18 groups of entrepreneurships situated in eight different regions with the biggest probability of clusters existence.[22] The first attempt at diagnosing and mapping potential clusters in Poland was undertaken by IBnGR, after entering the European Union structures during the years 2004 and 2005. Further research was conducted by the Polish Agency for Enterprise Development (PAED).

Since that time various programmes and initiatives (with a focus on clusters) have been implemented in Poland, both on the EU and national levels.

---

[21] Innovating Regions in Europe. http://www.innovating-regions.org/download/Clusters_in_CCE.pdf (accessed February 7, 2009).

[22] http://www.klastry-efs.pl/index.php?option=com_content&task=view&id=20&Itemid=40 (accessed February 7, 2009).

The aim of all the programmes, documents, projects and initiatives related to the cluster concept is to support clusters and cluster initiatives development in the country. However, due to the fact that there is no defined cluster policy in Poland, all the sources of support are scattered within different policies, such as the innovation policy or the policy focused on regional development.

Nine strategies have been prepared as a basis for a new strategic layout for management of the country's development policy. The policy concept of economic development based on clusters is embedded in the "Strategy of innovation and economic efficiency" developed by the Ministry of Economy[23].

Numerous terminologies defining clusters, for example, cluster, bunch, co-operation network, economic cluster, entrepreneurship cluster, industrial complex, industrial bundle, industry cluster or local production system exist in Poland (Jagódka 2009, Pilarska 2009).

In Polish legislation, the definition of a cluster appears for the first time in paragraph 13, Act 2 in the Regulation of the Minister of Economy from 2nd December 2006 on the provision of financial assistance non-related to the operational programmes by PAED (Journal from 2006 No. 226, item: 1651 with further changes). According to this definition, clusters should be understood as a spatial and sectoral concentration of at least ten entities operating for economic development or innovation, on the territory of one or a few neighbouring voivodeships, which compete and co-operate with one another in the same or related branches and are connected within a developed network of co-operation of a formal or informal character, where at least half of the entities consist of entrepreneurs.[24]

Clusters - supporting authorities and institutions

There are **authorities on a ministerial level** responsible for implementing a cluster-based policy and dealing with issues concerning co-operation in Poland, such as:

a) the Ministry of Economy,
b) the Ministry of Regional Development,
c) and the Ministry of Science and Higher Education.

---

[23] "Kształtowanie polityki klastrowej w Polsce", http://www.mg.gov.pl/node/11789 (accessed May 30, 2011).

[24] "Rozporządzenie Ministra Gospodarki z dnia 2 grudnia 2006 r. w sprawie udzielania przez Polską Agencję Rozwoju Przedsiębiorczości pomocy finansowej niezwiązanej z programami operacyjnymi". http://isap.sejm.gov.pl/DetailsServlet?id=WDU20062261651 (accessed May 7, 2011).

PAED is subordinate to the Ministry of Economy and appointed as the main agency supporting clusters. It is responsible for implementing a cluster-based policy on a national level. There are also additional institutions dealing with cluster concept development, for example IBnGR, the Centre for Social and Economic Research, the Polish Foundation for SME Development and Promotion, or the National Centre for Research and Development.

The cluster-based policy framework in Poland mainly lies in the hands of the Ministry of Economy and **PAED**, which plays a supportive role towards clusters and at the same time leaves the idea of cluster creation up to the companies themselves. PAED offers numerous activities in regard to clusters support (for example mediation in financial support, creation of a business-friendly environment for companies, supporting business surrounding institutions) and participates in cluster policy creation on a government level (Kryjom 2008).

Another institution, which plays a significant role in cluster concept development, is the **Gdańsk Institute for Market Economics** which has been involved in cluster issues from the very beginning. The engagement and development of interest in the cluster concept naturally developed in connection with the economic policy in the country. At the beginning, IBnGR took part in developing numerous pilot projects and supporting cluster initiatives in Pomerania. It has consequently been assisting in supporting initiatives throughout Poland. At present, IBnGR acts as a think-tank, which together with PAED, aims at establishing a cluster policy in the country.

Currently the Agency, in co-operation with the Institute, undertakes a project focused on Polish clusters and cluster policy aiming to strengthen clusters, increase their competitiveness and innovative ability through the development of human capital and a cluster policy.

In addition, there are **actors on a regional level** (for example Regional Development Agencies, Marshal Offices in particular voivodeships), which support the activities of SMEs, innovativeness development, promote the regions, and are responsible for economic and social stimulation (they usually indirectly support cluster development). The type of steps undertaken and the level of their involvement in cluster issues vary throughout the regions, from those providing a wide spectrum of support starting from serving as an intermediary in providing financial support, through providing information support, training of instructors/facilitators, promotion of innovation, active participation in the formation of clusters, seeking companies suitable for cluster formation, educational activities in the field of clusters and innovations, up to help in preparation and management of projects (for example Warmia and Mazury Regional Development Agency in Olsztyn), to institutions that indicate no support of clusters at all. The conducted research of the Marshal Offices (11 out of 16 existing, 69% of the return) from the particular voivodeships in Poland provided a fresh look at the intensity of support of the regional authorities on cluster concept development. The majority of the Marshal Offices began to be engaged in the clustering idea from either 2006 or 2007. The youngest in the sample was the Marshal Office of Świętokrzyskie Voivodeship in Kielce, which has been involved in cluster issues since 2010. The majority of the Marshal Offices in the sample provide promotion of innovation (89%) and information support (78%) within their activities. The role of an intermediary in providing financial support, promoting clusters and active participation in the clusters formation (67% each) are also popular.

Additionally, the need for the establishment of a professional association devoted to the concept of cluster development in Poland has been noticed. Therefore, in 2010, the **Procluster** - Association of Professionals and Experts in Networks and Clusters was founded, as an association with headquarters in the Silesian Voivodeship, bringing together professionals and experts on networking and clustering.

Only recently, further initiatives have been established, such as:

- "Clustering Poland" Association focused on active building, strengthening and developing of Polish co-operation relationships, networking and cluster organisations on local, regional, national and international levels, as well as promotion of the clustering idea.
- Polish Clusters Employers' Association, which, as a platform uniting the Polish cluster environment, is representing the interests of the co-ordinators of clusters and cluster initiatives, protecting their rights, and creating favourable conditions for the improvement of their operation and competitiveness.
- The Wielkopolska Clustering Centre oriented on implementation of the Wielkopolska model of a cluster ecosystem in the region, strengthening relations with experts and the Centre's partners, as well as with national and regional administration, and focusing on the problems and objectives of the regional cluster environment.

An overview of government programmes of support for clusters

Despite the fact that a cluster-based policy was introduced in Poland in 2002, the programmes focused on supporting cluster development appeared several years later. After accession to the European Union, policy makers in the country began to place more attention on the issue of clustering and treated clusters as a tool stimulating effective innovation and competitiveness development.

Various programmes and initiatives (with a focus on clusters) have been implemented in Poland, both on **the EU and national level**.

On the EU level, the **INTERREG IIIC** programme, which ran from 2002 untl 2006, enabled large-scale information exchanges and experience sharing, thus playing a significant role in the development of networks of co-operation (bringing benefits such as experience in seeing what a cluster is, development of technologies substantial for a particular industry, etc.). The promotion of interregional co-operation is being followed by the INTERREG IVC launched in 2007 for the time period 2007-2013.

One of the very first programmes initiated on a national level was the **Training Programme on Clustering** (2005-2007) announced by PAED. Its main goal was to introduce to both entrepreneurships and local government authorities a new idea of co-operation in the concept of clusters, to promote clustering (by training and organising conferences regarding clusters) and undertaking shared ventures vital for companies and regional development. Thanks to these activities, assistance in new cluster creation and the development of those that already existed were provided (Kosińska 2007).

Since the idea of co-operation within a cluster began to attract the desired attention of more and more groups of companies, new projects focused on supporting clusters (also financially) were developed.

In 2007, the **Pilot Programme: Support for Clusters** with the main goal of improving and developing the competitiveness of the existing cluster structure was implemented.

Programmes such as **BSR InnoNET**, **CEE-ClusterNetwork**, and **INNET** were implemented on the EU level (including Poland) within the time frame of 2006-2009. They aimed at creating an environment for building strong industrial clusters and innovation poles in order to focus on cluster and network policy and to promote interaction and co-operation between regional, national and European funding schemes.

In the meantime, the **Innovation Express** (IE) programme was announced in 2008. It based its main objective on activities devoted to initiating, developing and boosting technological co-operation among European clusters, enhancing trans-national co-operation and aiming for future joint R&D and innovation project development. IE was implemented for two years and offered a special "matching tool" as an opportunity for clusters to establish contacts with new partners (other clusters) for future trans-national collaboration.[25]

Due to major interest in activities regarding the internationalisation of clusters, the idea of the IE programme has been continued within a new component of "market expansion" within the Operational Programme Innovative Economy 2007-2013 (OP IE) Priority 5.1 with a similar area of aggravated activities for support.

In 2008, the pilot programme **Bonds for Innovation** was also announced and continued up until 2011. The Programme aimed at financially supporting micro and small enterprises, initiating contacts amongst them and research institutions (thus indirectly supporting clusters).

In Poland, the majority of the initiatives are undertaken thanks to support from the government, including European Union resources. Shared governmental and company resources, however, are also utilised quite often.

Clusters were additionally implementing projects under the Foresight Programme (2003-2006) and the European Territorial Co-operation Programme (ETC) 2007-2013. Also, implemented on the EU level were the Leonardo da Vinci Programme, the CIP, and the 7th Framework Programme for Research, Technological Development and Demonstration (FP7-RTD) which through supporting education, innovations and competitiveness influence cluster concept development in Poland.

Apart from activities on a national level, certain steps were also undertaken on a **regional level**. A substantial instrument of support was the Integrated Regional Operational Programme 2004-2006 (IROP) aimed at regional innovativeness growth. It focused on co-operation development among R&D and the private sectors and hence indirectly supported clusters progress[26]. This programme was complementary to another, the Sectoral Operational Programme "Improvement of the Competitiveness of Enterprises, years 2004-2006", which amongst other things focused its attention on strengthening the business support for institutions as well as co-operation between the R&D sphere and the economy, and the creation of favourable conditions for company development. In addition, the Sectoral Operational Programme "Human Resource Development for the years 2004-2006" with its Measure 2.3 "Development of personnel in a modern economy" contributed to clusters and cluster initiatives support

---

[25] www.proinno-europe.eu html (accessed May 5, 2011).

[26] http://www.funduszestrukturalne.gov.pl/NR/rdonlyres/718458E5-53A5-4486-9CE2-657036614EBE/0/irop_pro.pdf html (accessed May 5, 2011).

as within that measure a training programme on clustering was implemented.

From the time the importance of the cluster phenomena was highlighted by the European Commission for the period 2007-2013, the National Strategic Reference Frameworks (NSFRs) and Operational Programmes, in particular European countries (also in Poland) have devoted more attention to clusters. The main strategic goal of the NSRF 2007-2013 in Poland is to create conditions for an increase in the competitiveness of the Polish economy based on knowledge and entrepreneurship. This goal is being realised through more specific aims pursued within the so-called operational programmes administrated by the Ministry of Regional Development, regional operational programmes administrated by boards of particular voivodeships and projects co-financed from structural instruments.

A special mention of the **Operational Programme Innovative Economy** for the years 2007-2013 should be made. The programme's main goal is the development of the Polish economy based on innovation entrepreneurships. Among the detailed goals are an increase of innovation of entrepreneurships, competitiveness of Polish science and the use of information and communication technologies in the economy and many others. There are nine priorities inside the programme, where the **5th Priority "Diffusion of innovations"** is primarily connected with clusters and co-operation. The main goal of this priority is to assure the high quality of service and an infrastructure for entrepreneurships, which may help to strengthen their potential for innovation, and in addition, assure the strengthening of the entrepreneurship competitiveness position by development of co-operative connections (Ministerstwo Rozwoju Regionalnego 2007, www.parp.gov.pl). The priority underlines that co-operation among entrepreneurships and between entrepreneurships and surrounding institutions, in particular research units, plays a significant role in their development and creates convenient conditions for the creation of new solutions, the exchange of experience and better use of the resources owned by individual entities. The largest component of support for cluster development lies in Measure - 5.1 "Support for the development of supra-regional co-operative relations", under which the support for joint projects of the groups of participants is being granted. In addition, Measure 5.2 "Support for pro-innovation networks of business support institutions with a supra-regional meaning" within which pro-innovative business support institutions such as science and technology parks, technology incubators, technology transfer centres, etc. may be supported, has a significant impact (Ministerstwo Rozwoju Regionalnego

2007). In order to obtain financial support, however, within this programme, specifically from Measure 5.1, the applicants must attain certain conditions[27,28,29].

The necessity for supporting co-operation and development of co-operation connections in Poland was also articulated in the strategic document "Directions for an increase in economic innovation for the years 2007-2013". The OP IE constitutes one of the instruments implementing the steps of this document accepted by the government in 2006 (articulating in the chapter "Infrastructure for Innovation" the significance of supporting joint activities of enterprises with a networking character and directed at the implementation of innovation ventures)[30].

Additional operational programmes influencing cluster development are **Operational Programme Human Capital** (OP HC) 2007-2013 with its Priority II and Sub-measure 2.1.3. System support for increasing the adaptation potential of employees and entrepreneurs, and **Operational Programme Development of Eastern Poland** (OP DEP) with its Priority I: Modern Economy and Sub-measure I.4., Promotion and co-operation. Within OP DEP component 1.4.2., Co-operation - establishing and developing clusters, attention is being paid to building networks of co-operation regarding innovative economic activity, for example, identification and development of clusters[31,32]. Within OP HC Sub-measure 2.1.3., PAED conducted within the time frame of March to

---

[27] "Guide Book. 13 ways to become innovative. Establishing a successful business within the Operational Programme Innovative Economy", 2009. Warsaw: PAED.

[28] "Instrumenty wspierania klastrów", http://www.mg.gov.pl/Wspieranie+przedsiebiorczosci/Wsparcie+finansowe+i+inwestycjeKlastry/Instrumenty+wspierania+klastrow (accessed May 30, 2011).

[29] "Przewodnik po kryteriach wyboru finansowanych operacji w ramach Programu Operacyjnego Innowacyjna Gospodarka, 2007-2013", 2008. Warszawa: Ministerstwo Rozwoju Regionalnego. http://www.konkurencyjnosc.gov.pl/NR/rdonlyres/2DBE47EF-384E-4986-9E5C3FDE44DBCF66/46624/Przewodnik_po_kryteriach_200509.pdf (accessed May 5, 2011).

[30] "Kierunki zwiększania innowacyjności gospodarki na lata 2007-2013", http://www.mg.gov.pl/NR/rdonlyres/90AF42C4-A420-4BF9-9CE8-08C28B8E4FFE/50360/KierunkiPL.pdf (accessed February 7, 2009).

[31] "Operational Programme Development of Eastern Poland 2007 – 2013". Warsaw: Ministry of Regional Development, October 2, 2007. http://www.mrr.gov.pl/english/european_funds_2007_2013/european_funds_2007_2013/PORPW/Documents/PORPW_eng_final.pdf (accessed February 28, 2011).

[32] "Broszura informacyjna Program Operacyjny Rozwój Polski Wschodniej". Ministerstwo Rozwoju Regionalnego, http://www.polskawschodnia.gov.pl/NaborWnioskow/Documents/PORPW.pdf (accessed April 6, 2011).

November 2010 a project entitled "Benchmarking of clusters in Poland". Its aim was to deepen awareness and knowledge regarding the status of cluster development in Poland and promote the idea of clustering. In addition, the project was intended to deliver information about the possibilities for improvement of their own steps of clusters and show the best practices from Polish clusters. Furthermore, it also aimed at an indication of recommendations regarding cluster policy development in Poland directed to the authorities and other institutions supporting and co-operating with clusters (PARP 2010). In 2012, the second edition of this benchmarking project was conducted (PARP 2012). In addition, under OP HC sub-measure 2.1.3, a project prepared by PAED and performed as part of the system project entitled "Development of human resources through promotion of knowledge, transfer and popularisation of innovation" was implemented (2011-2012). Its main activities were focused on supporting cluster development in Poland and supporting institutions creating a cluster policy.

PAED also participated in the project entitled **"NGP Excellence clusters – Cluster excellence in the Nordic Countries, Germany and Poland"** focused on benchmarking of clusters and cluster policies in seven countries including Poland. The project aimed at promoting cluster management excellence and mutual learning by comparing both cluster organisations and cluster support programmes in the specified countries[33].

In addition, the **Regional Operational Programmes** (ROPs) for the years 2007-2013 are being implemented in particular voivodeships providing fundamental strategic solutions at the regional level. Their objectives are in compliance with regional development strategies and with the goals of NSFR, focusing on enhancing the competitiveness of individual regions and promoting balanced development. Their content is harmonised with the specifics of the region since one of the primary goals of ROPs is to support entrepreneurship development and growth of an innovative economy i.e. an increase in regional competitiveness. In order to achieve these goals, various activities (financial and institutional support, consultancy and promotion) directed at business entities especially SMEs are being designed. Stimulating activities focused on creation and/or development of networks of co-operation (new organisational solutions) among research units, educational units, business units and entrepreneurs (building a lasting bridge between science and business), and competitiveness of entrepreneurships, which constitute an

[33] "The NGPExcellence Project. Cluster Excellence in the Nordic Countries, Germany and Poland", http://www.pi.gov.pl/PARPFiles/file/klastry/Leaflet_NGPE.pdf (accessed May 5, 2011).

inseparable component of clusters/cluster initiatives growth in the regions, appears as an integral element. A number of ROPs clearly specify the fact of providing assistance for the development of co-operative relations i.e. creation of clusters and/or support of projects in the field of cluster management, initiating and animating co-operative relations including clusters, etc. (for example the Regional Operational Programme for Lesser Poland[34], Regional Operational Programme for the Warmian-Masurian Voivodeship[35], Regional Operational Programme for the Kuyavian-Pomeranian Voivodeship 2007-2013[36], Regional Operational Programme for the Masovian Voivodeship 2007-2013[37])[38]. Half of the ROPs in Poland place an increase in innovativeness and support of entrepreneurship in first place, as well as strengthening the competitiveness of enterprises in the region. Another third allocates this priority to second position in significance, just after the improvements in transport, the regional and local infrastructure. The Pomeranian Voivodeship, for example, implemented in addition the Regional Programme for Cluster Support for the Pomeranian Voivodeship for the years 2009-2015, aimed at strengthening the competitiveness of the economy by stimulating the creation and development of regional and local clusters in this voivodeship.

Additionally, each voivodeship has its own **Regional Innovation Strategy** (RIS) focused on optimising the regional infrastructure (its compatibility with the needs of SMEs), narrowing down the weak points of the region and stimulating the transfer of information and knowledge into the sub-regions, thus having an indirect influence on clustering.

Although the issue of clusters is gradually growing in popularity and slowly appearing in regional development strategies, a continued lack of

---

[34] "Małopolski Regionalny Program Operacyjny na lata 2007-2013". Zarząd Województwa Małopolskiego, http://www.malopolskie.pl/Pliki/2007/MRPO_2007 1004.pdf (accessed February 28, 2011).

[35] "Regionalny Program Operacyjny Warmia i Mazury na lata 2007-2013". Zarząd Województwa Warmińsko – Mazurskiego, http://www.rpo.warmia.mazury.pl/ index.php?page=dzial&dzial_id=79 (accessed February 28, 2011).

[36] "Regionalny Program Operacyjny Województwa Kujawsko-Pomorskiego na lata 2007-2013". Zarząd Województwa Kujawsko-Pomorskiego, http://www.kujawsko-pomorskie.pl/files/roz_reg/prog_oper/rpo/RPO_2007-2013_WERSJA_IV.pdf (accessed February 28, 2011).

[37] "Regionalny Program Operacyjny Województwa Mazowieckiego 2007-2013". Urząd Marszałkowski Województwa Mazowieckiego, http://rpo.mazowia.eu/data/other/rpo_1.pdf (accessed February 28, 2011).

[38] On the basis of all the documents of Regional Operational Programmes in Poland.

importance assigned to it leaves it as a neglected area, which requires further attention.

The current cluster landscape in Poland

The cluster phenomenon in Poland began to grow in popularity over the last decade. Despite the fact that interest in cluster structures is fairly new in this country, increasing numbers of clusters and cluster initiatives emerge every year.

According to Portal Innowacji (PI)[39], PAED identified 130 clusters and cluster initiatives in Poland as of November 2008. Since that time, the cluster landscape has been enriched even more. Map 2-1 presents the dispersion of 215 clusters/cluster initiatives in Poland based on the voivodeships with the status as of 2012.

The latest research "Benchmarking of clusters in Poland - 2010" identifies 178 clusters/cluster initiatives in Poland in 2010 and over 250 clusters/cluster initiatives in 2012 (PARP 2010 and 2012). An accurate number is difficult to define, however, as many of them are listed by official sources, for example PAED, but in fact are either in a re-organisation process or have stopped (stagnated) their activity or simply do not exist any longer.

In general, the highest density appears in central-eastern and southern (apart from Opole Voivodeship) parts of Poland; with the largest number of clusters/cluster initiatives existing in the Masovian, Podlaskie and Lesser Poland voivodeships (respectively 32, 16 and 16 entities). Also, Greater Poland voivodeship has a high number of 23 clusters. The lowest number appears in Opole and the Pomeranian voivodeships (respectively 3 and 6 entities).

The process of clustering occurs in all the Polish regions. A strong potential for competitive cluster development can be distinguished. Particular branches of Polish clusters and cluster initiatives are shown in Map 2-2.

[39] http://www.pi.gov.pl/PARP/data/klastry/index.html (accessed January 25, 2011) ©PARP.

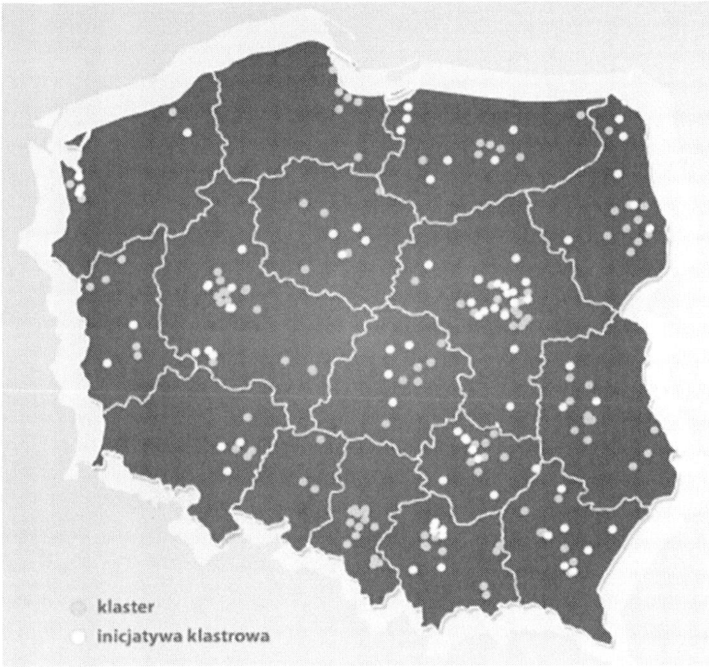

Map 2-1 Map of clusters and cluster initiatives in Poland based on voivodeships
Source: Adapted from http://www.pi.gov.pl/PARP/data/klastry/index.html ©PARP (accessed February 25, 2013), map prepared on the basis of data from November 2008, updated in 2012

The versatility of the branches in which the Polish clusters exist is recognised. Strong cluster structures are observed both in traditional and high technology branches. Within high-tech industry branches, for example the aviation industry, automotive industry, and automation, space technologies can be distinguished, along with informatics and communication technologies (for example information systems, media, advertising), as well as biotechnology or life sciences. Among the traditional branches, the food industry (for example ecological food, food-processing and brewery industry), chemical and wood industries (for example furniture production) can be identified.

Clusters in Poland are characterised by an early stage of development. The majority of them emerged in 2006 and 2007 or later.

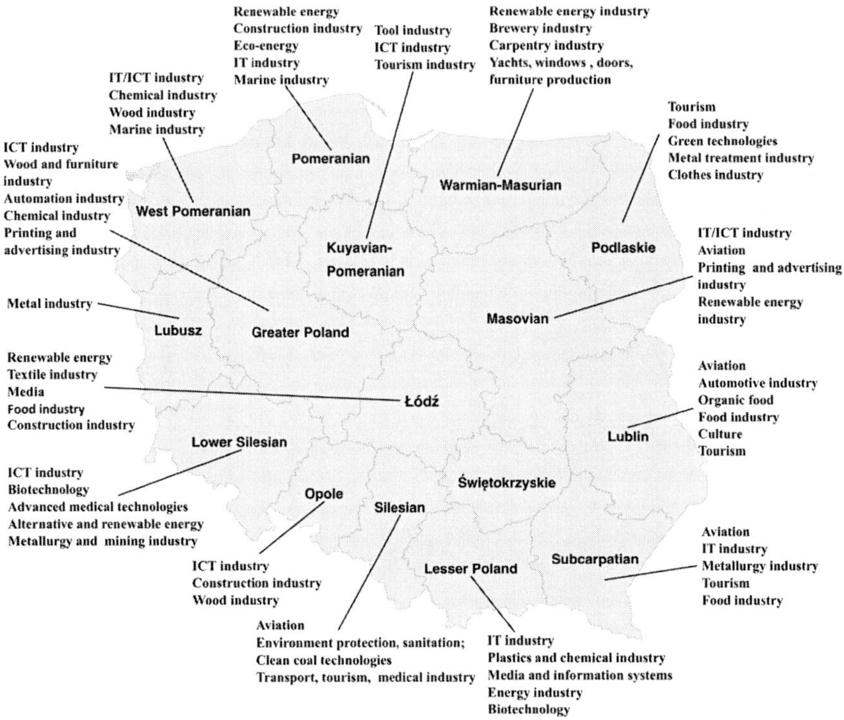

Map 2-2 Division of predominant industries of Polish clusters/cluster initiatives based on voivodeships
Source: Own development

Cluster policy in the Czech Republic

The beginnings of a development of a cluster policy in the Czech Republic have been dated to the year 2000. The very first activities aimed at development of co-operation were undertaken in 2001 with the implementation of the CO-OPERATION Programme for the years 2001-2004. At that time, attention was devoted to the expansion of networking among SMEs. At a later point, steps towards supporting clusters began to be initiated. However, the vivid development of activities appeared with accession to the European Union (Poledna 2007).

A promising beginning for the development of a cluster policy in the country came about with the acknowledgement and implementation of the

National Cluster Strategy 2005-2008 by the Czech Government. The Strategy constituted the main document on clusters in the Czech Republic. In its objectives, it placed an emphasis on the usage of clusters in order to interconnect the resources and programme measures under various strategies and policies. Additionally, it pointed out the importance of the development of a dialogue with regions, tertiary education institutions and the private sector as well as the issue of helping SME groups work together.

Simultaneously with the National Cluster Strategy, the attention of the Operational Programme Industry and Enterprise (OPIE) for the years 2004-2006 has been focused on the issue of clusters. Within the so-called CLUSTERS Programme, support explicitly focused on clusters and cluster initiatives has been provided, thus stimulating their development in the country. In addition, the identification of clusters and sectors for potential cluster emergence was carried out under the supervision of the Investment and Business Development Agency – CzechInvest.

Although support for clusters has been continued within the on-going Operational Programme Enterprise and Innovation (OPEI), no further steps have been undertaken towards a conceptual strategy for cluster policy development in the Czech Republic.

The Czech equivalent for the term cluster is being widely used in the country.

Cluster - supporting authorities and institutions

The **Ministry of Industry and Trade** (MIT) and their subordinate **CzechInvest** can be distinguished among the actors responsible for the co-ordination of activities and programmes focused on clusters in the Czech Republic. The MIT is responsible for the conceptual side of the cluster phenomenon and cluster policy implementation in the country, whereas CzechInvest is in charge of its practical application.

CzechInvest agency focuses its activities on strengthening the competitiveness of the Czech economy through supporting SMEs, business infrastructure, innovations and attracting foreign investments in the areas of manufacturing, business services and technology centres. It is widely involved in cluster issues, both in the application of a policy based on clusters as well as being the Intermediate Body assisting in providing support for cluster organisations. The agency fosters the establishment and development of not only clusters, but also other co-operative groups such as poles of excellence or technology platforms, where the actors from different pillars of a Triple Helix collaborate with one another. During the first wave of a cluster supporting scheme, CzechInvest provided a wide range of assistance, not only in information and financial support, but also in the form of wide publicity, as well as in awareness and capacity building-education and certification of facilitators, etc. In 2004, it drew attention to the phenomenon of clusters via offering an educational programme explicitly focused on clusters. The programme gathered cluster facilitators from throughout the country, academics from tertiary institutions, representatives of regional governments and regional institutions and the private sector providing them with training and familiarisation with the cluster concept. The agency currently provides financial support for cluster organisations within the on-going OPEI. Applicants, however, have to meet certain requirements.

The initial euphoria and enthusiasm about a cluster oriented policy and initiatives undertaken on a national level has been weakened over time. Additionally, the marginal involvement of regional policy actors has been observed.

The **National Cluster Association** (NCA) was consequently founded as a long-term and competent platform for cluster development in the country. It has been actively involved in cluster issues since 2010. NCA aims at bringing "together organisations and individuals" and "on the basis of concentration of knowledge, experience and expertise" ensuring the sustainable development of cluster initiatives and cluster policy in the country, thus, strengthening its competitiveness[40]. The association focuses on representing the interests of Czech cluster initiatives, facilitating their development and stimulating their internationalisation (representing them on an international arena). It provides information support, training of instructors/facilitators, promotes clustering, is directly (actively) involved in the establishment of clusters and recruits companies suitable for clustering. NCA associates not only clusters, but also universities, regional and innovation agencies, and consultants.

The **actors on a regional level** are rather marginally involved in cluster issues, if involved at all. They are mainly focused on the co-ordination of a development policy. Worthy of a mention as an exception are the Union for the Development of the Moravian-Silesian Region, which supported the mapping, establishment and promotion of industry clusters contributed to the formation of regional industry clusters, and the Regional Development Agency Ostrava, which is supporting clusters through its ClusterNet initiative.

---

[40] www.nca.cz

An overview of government programmes of support for clusters

Active support for clusters development has existed for several years. The **CO-OPERATION Programme** was the first elaborated by MIT with an aim of supporting the policy for SMEs and linking cluster development. The first activities directed at cluster support began to develop.

Several years later, the implementation of the **National Cluster Strategy** constituted a major step in cluster concept development in the Czech Republic. This document contained the main principles, measures and aims of the application of a successful economic model of a cluster in regard to the conditions existing in the country. It highlighted the significance of competitiveness, co-operation and innovation, and stressed support for SME development. The Strategy pointed at the use of clusters in order to link the resources of several ministries and regions in the Czech Republic and integrate strategies and policies regarding the development of SMEs, innovation, R&D, exports, education and/or infrastructure. Furthermore, it aimed at creating a framework for analysis, monitoring and evaluation of the performance of cluster initiatives and their impact on the regional and national economy, which would also enable the benchmarking of these initiatives within and outside the country[41]. In order to fulfil the targeted goals, measures such as support for cluster concept (development), support for cluster facilitators and managers, mapping of clusters, development of clusters and measuring the performance of cluster initiatives were distinguished (Neužilová 2006).

Subsequently, within the OPIE[42] and its 1st Priority "Development of business environment" the **CLUSTERS Programme** was implemented for the programming period of 2004-2006, placing an emphasis on clusters as a tool for stimulating competitiveness and national growth, and on strengthening the communication and co-operation links among enterprises (the business sphere) and science and research spheres (Innovating Regions in Europe, n.d.; OECD 2007). It constituted the first programme explicitly focused on clusters and their development in the country and was carried out under the patronage of CzechInvest. The Programme focused on two phases. The first mapping phase was devoted to the identification of existing potential and searching for suitable companies for clusters (a profound analysis of the given sector and the

---

41 "Cíle Národní klastrové strategie 2005 – 2008". Vláda České Republiky. http://www.mpo.cz/dokument6216.html (accessed February 28, 2011).

42 "Operační Program Průmysl a Podnikání na léta 2004 – 2006". Ministerstvo Průmyslu a Obchodu České Republiky, http://www.evaluace.cz/dokumenty/oper_progr_cr/op_prumysl.pdf (accessed February 28, 2011).

facilitation of the actors towards co-operation within clusters), whereas the second phase focused on the establishment and development of the cluster. Within these two phases, the largest concentration of support was granted in the Moravian-Silesian, South Moravian and Hradec Králové regions.

Currently, support for clusters is being provided within the on-going OPEI and its **CO-OPERATION Programme**, within the Priority 5th Axis "Environment for Enterprise and Innovation". The Programme aims at creating a favourable entrepreneurial environment and supporting the formation and development of co-operation groups i.e. clusters and technology platforms. It concentrates on strengthening the innovative potential and the use of new technologies as well as aiming at stimulating co-operation between enterprises and research institutions[43]. This new programme was broadened in comparison with its predecessor, being less restricted in regard to the industry sphere. However, certain fields of support, for example for cluster mapping, are no longer provided. The Programme also promotes internationalisation through the **CORNET project** of the FP7.

Additional operational programmes have also been implemented within the time frame of 2007-2013, playing an important role in cluster development in the Czech Republic though indirectly influencing their growth. Among them the following can be distinguished: the Operational Programme Human Resources and Employment (OP HRE) focused on strengthening active labour market policies; the Operational Programme Research and Development for Innovation (OP R&DI) focusing on commercialisation and popularisation of R&D; and the Operational Programme Education for Competitiveness (OP EC) where special attention has been paid to the preparation of human resources for the formation and functioning of technologically focused clusters[44,45,46].

---

[43] "Operational Programme Enterprise and Innovation". Ministry of Industry and Trade of the Czech Republic, http://www.czechinvest.org/data/files/oppi-msc-en-29-11-schvalen-ek-674.pdf (accessed February 28, 2011).

[44] "Operational Programme Research and Development for Innovation". Ministry of Education, Youth and Sports, July 24, 2008. http://www.msmt.cz/strukturalni - fondy/operacni-program-vyzkum-a-vyvoj-pro-inovace-1 (accessed May 23, 2011).

[45] "Operational Programme Education for Competitiveness". Ministry of Education, Youth and Sports, http://www.strukturalni-fondy.cz/getdoc/5257c95e-fac7-45bb-8313-a489678b9010/Dokumenty (accessed May 23, 2011).

[46] "Operational Programme Human Resources and Employment 2007-2013". Ministry of Labour and Social Affairs, http://www.strukturalni-fondy.cz/getdoc /61e1c103-49f7-4d5a-b809-1142bce99472/Dokumenty (accessed February 28, 2011).

In addition to this, strategic documents such as the National Innovation Policy of the Czech Republic for the years 2005-2010, Strategies for Regional Development in the Czech Republic for the years 2007-2013 or the National Strategic Reference Framework of the Czech Republic 2007-2013 play a supportive role in cluster development.

Apart from programmes undertaken on a national level, the policy regarding clusters is also being pursued on a regional level within the **Regional Operational Programmes** (ROP) and **Regional Innovation Strategies** (RIS). Through covering several thematic areas with the aim of accelerating development and increasing regional attractiveness for investors as well as facilitating innovation and infrastructure development, promoting entrepreneurship and creating favourable supportive conditions for enterprises, they indirectly support cluster development. The differences in the importance and intensity of the steps undertaken with regard to clusters are visible within the regions. A number of them appear as active, while others treat the cluster issues marginally.

The Pardubice Region, for example, introduced within its RIS a specific measure called "Development of clusters and co-operation networks", pursuing concrete activities aimed at increasing awareness of the usefulness and convenience of the formation of clusters; financial, methodological and information stimulation of creation and development of clusters in the region and searching for trans-regional clusters and involving firms in these clusters[47]. The Pardubice Regional Development Agency also supported a project focused on "Searching for Companies suitable for the Creation of an Electrotechnical Cluster in the Pardubice Region"[48]. Support for the functioning of existing and newly established clusters in the Zlin Region, active support and co-operation in the preparation and implementation of innovative cluster projects, searching for further ways of support for clusters, and developing and implementing projects focused on mapping the possibility for new cluster creation all rank among those activities implemented within the specified measure. Moreover, since 2005, the Technology Innovation Centre (TIC) in Zlin has actively participated in mapping the possibility of cluster establishment, and also in promoting the emergence and development of new clusters in the region[49]. The Moravian-Silesian Region within its strategy highlights

---

[47] "Regionální inovační strategie Pardubického kraje". Regionální rozvojová agentura Pardubického kraje, http://www.pardubickykraj.cz/article.asp?thema=4011&item=55061 (accessed May 5, 2011).

[48] http://www.elektroklastr.cz/ (accessed May 5, 2011).

[49] "Regionální inovační strategie Zlínského kraje", 2007. http://ris.utb.cz/ (accessed May 5, 2011).

the importance of initiation and implementation of cluster projects in R&D area[50].

Other regions also underline in their regional documents the significance of clusters as spatial forms of co-operation, platforms for long-term co-operation between firms (the business community) and academic institutions, platforms for use of R&D results, development of co-operation with practice in teaching and R&D activities, etc. The concepts of linking businesses, initiation, organisation and support in business networking, communication and co-operation, as well as mapping and supporting cluster development in regions slowly appears within various priorities and measures[51,52,53,54]. Nonetheless, the issue of clusters can still be considered a neglected area, which requires further attention.

The current cluster landscape in the Czech Republic

The issue of cluster development in the Czech Republic was analysed for the first time in the Moravian-Silesian Region, where a feasibility study focused on identifying the industry groupings was implemented in 2002. Since that time, awareness of the clustering idea has begun to grow and certain activities in regard to clusters/cluster initiatives development in the country have been undertaken. Altogether 79 clusters/cluster initiatives have appeared in the country (status as of November 2012), some of them were not established in the end and some declined or went through a re-organisation process after a certain period of their existence. In approximately 80%, a cluster with its own cluster organisation was created.

At the end of 2008, according to CzechInvest, there were 33 clusters and several cluster initiatives in the Czech Republic.

---

[50] Regionální Inovační Strategie Moravskoslezského kraje 2010 – 2016". Agentura pro regionální rozvoj, http://verejna-sprava.kr-moravskoslezsky.cz/assets/rozvoj_kraje/rk_ris_2010-16.pdf (accessed May 5, 2011).

[51] "Regionální inovační strategie Jihočeského kraje". Jihočeská agentura pro podporu inovačního podnikání, http://www.kraj-jihocesky.eu/index.php?par[id_v]=1545&par[lang]=CS (accessed May 5, 2011).

[52] "Regionální inovační strategie Jihomoravského kraje". Jihomoravské inovační centrum, http://www.kr-jihomoravsky.cz/Default.aspx?pubid=234&TypeID=7&foldid=2928&foldtype=7 (accessed May 5, 2011).

[53] "Regionální inovační strategie Královéhradeckého Kraje". Centrum EP - Regionální rozvojová agentura, http://www.kr-kralovehradecky.cz/assets/rozvoj-kraje/RVVI/Strategie-_rijen09.pdf (accessed April 5, 2011).

[54] "Regionální inovační strategie Libereckého kraje". ARR - Agentura regionálního rozvoje, http://www.portal-inovace.cz/userdata/files/RISLK_plna_verze.pdf (accessed May 5, 2011).

A variety of industry sectors, both in traditional as well as in high-tech branches, can be observed (Map 2-3), with a predominance of manufacturing industries.

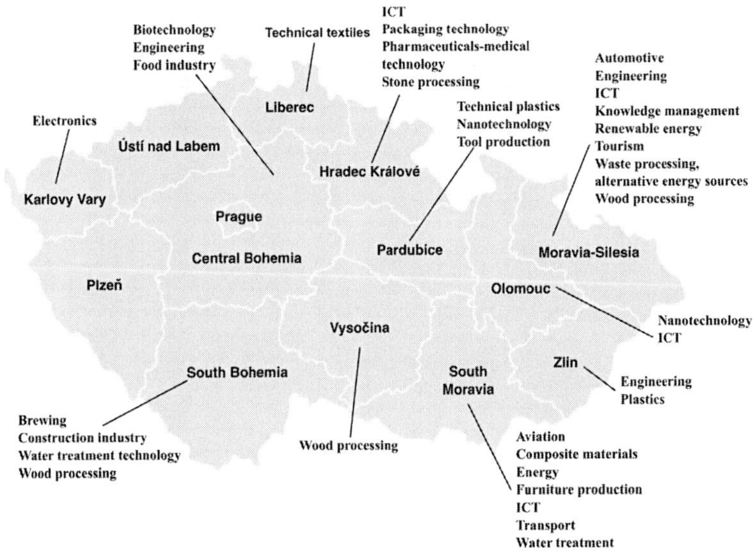

Map 2-3 Division of predominant industries of Czech clusters/cluster initiatives based on the regions
Source: Own development

Currently, more than 40 active clusters/cluster initiatives can be identified (status as of November 2012). The remaining number of clusters/cluster initiatives can be characterised as emerging or stagnating. The largest number of clusters exist in the Moravian-Silesian (14), South Moravian (12) and South Bohemia regions (11), while in contrast, there are regions such as Ústí nad Labem (2) or Plzeň (1) with poor interest in clusters/cluster initiatives development.

Clusters in the Czech Republic are characterised by their short period of existence; the oldest Czech Machinery Cluster emerged in 2003. Rapid development of clusters has been observed since 2006 in relation to the announcement of the subsidy programme OPIE Clusters (Figure 2-3).

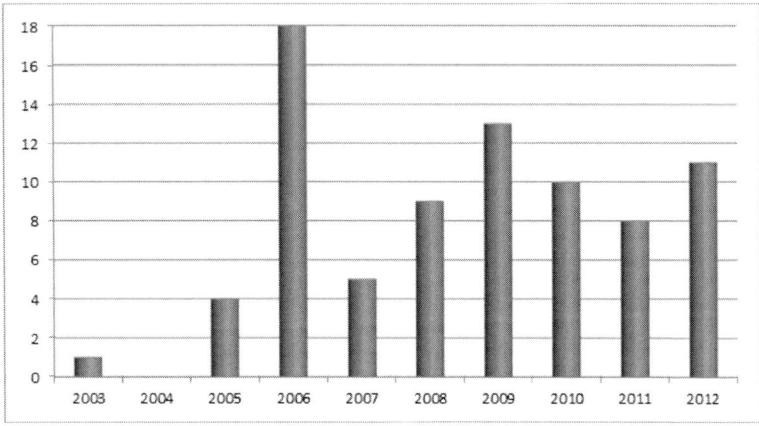

Fig. 2-3 Chronology of Czech clusters development
Source: Own development

Over half of all founded clusters utilised some form of subsidy (from either the OPIE or the OPEI programme).

Within the framework of support of the Programme Clusters (OPIE), a total of 53 projects have been supported, of which 41 projects concerned mapping and 12 projects concerned creating and developing a cluster. Subsidy pay-outs for projects reached a total of CZK 168.2 million, whereas CZK 27.7 million was spent on mapping and CZK 140.5 million on developing clusters. In general, it can be stated that 44 cluster initiatives were supported, of which the formation of a cluster organisation occurred in 30 cases.

An average of CZK 773,000 per cluster mapping project was requested for a company mapping (the maximum was CZK 1 million and the minimum was CZK 300,000). In terms of regions, the greatest support in mapping was received by clusters in the Moravian-Silesian Region and the South Moravian Region; a substantial portion of subsidies was also given to the South Bohemian Region and the Hradec Kralové Region.

The average request for establishing and developing clusters was CZK 16.5 million (the maximum was CZK 43 million and the minimum was just under CZK 5 million). In terms of regions, clusters in the Hradec Kralové Region received the greatest support; a substantial portion of subsidies was also given to the Moravian-Silesian Region and the South Moravian Region.

Today, clusters can draw support from the **Programme Co-operation (OPEI)**. During the first appeal, 17 projects were supported in 2009, in which clusters requested subsidies to a total amount of CZK 573 million. During the second appeal (with status as of May 2013), 24 projects have been approved, with requested subsidies totalling CZK 757 million.

In these appeals, clusters applied for an average of CZK 32.5 million (the maximum is CZK 80 million and the minimum is CZK 4 million). In terms of regions, in this period the most active clusters have been observed once again in the Moravian-Silesian Region and the South Moravian Region, however, recently also clusters in the Central Bohemian Region.

A comparison of the current and past programming period shows that more clusters exploited subsidy opportunities, and that they requested higher amounts of subsidies for their projects. Out of 12 clusters supported within the framework of the OPIE Programme, 7 clusters have once again applied for and received subsidies in the current programming period. In OPEI, a substantial portion of support was given to the newly formed clusters.

Cluster policy in the Slovak Republic

The Slovak Republic ranks amongst those countries with the shortest experience considering cluster concept development since it was the slowest to adopt the cluster-based policy in comparison with other EU countries - after the year 2005.

The first steps towards a cluster-based policy approach appeared only recently within the governmental documents for the time frame of 2007-2013. Up until that time, the cluster concept was considered marginally.

Due to this fact, no explicit cluster policy exists in the Slovak Republic. There is no organisation exclusively focused on supporting cluster development. Clusters are only treated as a means for regional development and support for SMEs. In addition, the unco-ordinated and fractured structure of associations, industrial associations and chambers of commerce, as well as the lack of a partner at the ministry level has delayed and/or hampered the development of clusters and cluster initiatives in the country.

The Slovak equivalent for the English term cluster has been used in the country.

Cluster - supporting authorities and institutions

There is **no ministry or agency explicitly focused on clusters** in the Slovak Republic. However, clusters primarily expect help from the Ministry of Economics, the Ministry of Education and the Ministry of Labour, Social Affairs and Family of the Slovak Republic, as well as from specialised agencies such as the National Agency for Development of SME's (NADSME), the Slovak Agency of Trade and Investments (SARIO) and the Slovak Energy and Innovation Agency (SIEA). Listed agencies only deal with clusters as a tool for support of SMEs and improving the business environment in the Slovak Republic. The agencies are expected to assist in the implementation of a cluster policy and support measures on a national level. However, thus far the issue of clusters is only marginal for all three agencies. A common means of support is information support. For example, in 2009, SIEA developed an analytical study entitled "Clusters as support for innovation development" and published a brochure "Clustering - Presumption of success" supporting in this manner clusters development in the country. SARIO, in turn, co-operates with a particular cluster. Its activity is focused on mutual assistance in co-operation on various events, exchange of professional information, participation in cluster project meetings, etc.

In the Slovak Republic clusters are being supported by higher territorial units and the private sector although their support is minimal. Higher territorial units often play the role of a "propagator" of the clustering idea and may implement activities such as training of facilitators, preparation of guidebooks, organisation of seminars and presentation of successful examples from abroad, etc.

The establishment of **the Union of Slovak Clusters** (UKS) in August 2010 was a huge step in the facilitation of clustering processes in the country. The Union was created by representatives of Slovak clusters with the aim of promoting the cluster phenomenon, and uniting and enforcing common objectives in the existing economic system in the country. It became a partner for communication in regard to cluster policy issues and clustering itself, both on the national and international level. Currently, it associates eight clusters with further acceding. The Union actively collaborates on the national level in the preparation of strategic materials for a cluster policy in the Slovak Republic. It is also a financial partner in the Central Europe Programme, which is its primary tool for start-up dialogues and encouraging the adoption of a cluster policy in the Slovak Republic. Its main goal is to support economic development and increase the competitiveness of all the Slovakian regions through the cluster phenomenon[55]. Furthermore, it constitutes a crucial "voice" of Slovak clusters, aiming not only to promote clustering in the country, but also to enforce implementation of the respective mechanisms, which would support cluster development in the Slovak Republic.

An overview of government programmes of support for clusters

In the Slovak Republic, the cluster concept appeared recently as part of the governmental documents such as: Innovation strategy of the Slovak Republic for 2007-2013 and **Operation Programme Competitiveness and Economic Growth** (OP C&EG) as part of the National Strategic Reference Framework 2007-2013 (Poledna 2007). The latter one is the first policy document mentioning clusters and their role in companies' competitiveness development.

Within the strategic Priority 2 "Knowledge economy" of the NSRF, the OP C&EG in the "Convergence" objective implements the specific priority 2.4 "Support of competitiveness of enterprises and services primarily through innovation". As stated in the NSRF document:

> Significant will be the support of co-operation between business and research and development institutions, universities and educational institutions and support for building the infrastructure for research, development and innovation, such as incubators, clusters, innovation centres, the centres of applied research, technology platforms, scientific technological and industrial parks, or regional innovation portals provided within that programme[56].

OP C&EG provides help in stimulating and increasing the competitiveness of enterprises and improvement of networking of SMEs within its measure "Innovation and technology transfers" and the linked measure "Support of innovation activities in enterprises", thus supporting clusters/cluster initiatives development[57].

The Operational Programme Research and Development has also a vital meaning. Through increasing the co-operation of R&D institutions, the creation and promotion of the transfer of newly acquired knowledge

---

[55] www.uks.sk (accessed May 6, 2011).

[56] "Národný strategický referenčný rámec 2007 – 2013". Ministerstvo Výstavby a Regionálneho Rozvoja Slovenskej Republiky, http://www.nsrr.sk/narodny-strategicky-referencny-ramec-2007-2013/ (accessed February 28, 2011).

[57] "Operational Programme Competitiveness and Economic Growth". Ministry of Economy of the Slovak Republic, http://www.nsrr.sk/en/operational-programmes/competitiveness-and-economic-growth/ (accessed February 28, 2011).

and technologies into practice provides support for the development of networks of collaboration.

In addition to this, the Operational Programme Employment and Social Inclusion within one of its measures provides comprehensive support for business development through implementing integrated system-designed projects with the aim of strengthening the innovative capabilities of enterprises by increasing the qualifications of their employees[58].

Attention is also being paid to co-operation amongst enterprises and cluster concept development in the Slovak Republic in the recent documents of Regional Innovation Strategies in the particular regions.

The Bratislava Region, for example, undertook an industry strategy to support cluster creation and development in the region within its RIS measure entitled "Creation of clusters in selected technology sectors"[59]. In addition, the Western Slovakia Region i.e. the Nitra Region[60], the Trnava Region[61] and the Trenčín Region[62] included activities focused on creating an environment conducive to development of innovative SMEs, their networking and clusters development in their RIS priorities and measures. For example, most of the activities of RIS in the Trenčín Region are aimed at SMEs and the regional government implements the strategy via administrative and financial support of organisations such as inter alia clusters. In the Trnava Region, RIS included a measure "The establishment and development of cluster of automotive suppliers and promotion of clusters in the relevant areas in the Trnava Self-governing Region" supporting the establishment and development of clusters in automotive and other prospective industries. Central Slovakia i.e. the Žilina and Banská Bystrica regions highlighted the importance of co-operation between industry and academia sectors, along with building

---

58 "Operačný Program Zamestnanosť a Sociálna Inklúzia". Ministerstvo Práce, Sociálnych Vecí a Rodiny Slovenskej Republiky, http://www.esf.gov.sk/documents/OP2007/OpZaSI_Final2007.pdf (accessed February 28, 2011).

59 "Regionálna inovačná stratégia Bratislavského samosprávneho kraja". 2004. Záverečná správa v rámci projektu RIS SK. BIC Bratislava, http://www.region-bsk.sk/SCRIPT/ViewFile.aspx?docid=56 (accessed May 5, 2011).

60 "Regionálna inovačná stratégia". Nitriansky Samosprávný Kraj. http://www.unsk.sk/files/regionalny/rissk_broz_sk.pdf (accessed May 6, 2011).

61 "Regionálna inovačná stratégia Trnavského samosprávneho kraja". http://www.trnava-vuc.sk/index.php/dokumenty_rr_sap.html?file=tl_files/dokumenty/dokumenty_odbor_strategie_a_programovania/dokumenty/ttsk/07-RIS_TRNAVA.pdf (accessed May 5, 2011).

62 "Regionálna inovačná stratégia pre Trenčiansky samosprávny kraj". RIS INSTIT, http://www.tsk.sk/buxus/docs//regional/ris/RIS_INSTIT_TSK.pdf (accessed May 6, 2011).

clusters in their RIS. In Eastern Slovakia i.e. the Košice and Prešov regions, cluster concept development is also meaningful. The Košice Region focuses on support of networking, clustering and development of innovation poles through combining SMEs in key and prospective sectors within its development strategy. Within the strategy, among other aspects, public support and assistance in networking of companies, searching for potential partners and use of shared services are being offered for cluster initiatives[63]. In the Prešov Region, within the activity entitled "Cluster creation support within the territory NUTS II - Eastern Slovakia" the strategy of supporting the establishment of clusters in machinery and robotics, tourism, health care and wood processing sectors is being implemented by the regional government[64].

Although enthusiasm in supporting cluster development has been expressed within recent regional documents in the Slovak Republic, a lack of activity in the implementation of respective steps can be observed. Further steps, therefore, need to be undertaken in order to stimulate clustering processes in the country.

The current cluster landscape in the Slovak Republic

The Slovak Republic has the shortest experience out of the presented Central European countries considering cluster concept development. At present, clusters in the Slovak Republic are in their emerging stage. Currently, an amount of around 18 cluster/cluster initiatives can be identified in the country. The Trnava region with the largest number of clusters/cluster initiatives (seven) is considered the leading region in clusters. However, a high interest in clustering processes is also visible in the Žilina, Banská Bystirca and Košice regions. Map 2-4 presents the division of predominant industry areas of identified clusters/cluster initiatives in the Slovak Republic based on regions. Three main areas can be distinguished: services–tourism sector, manufacturing industry and IT/ICT industry.

---

[63] "Regionálna inovačná stratégia Košického samosprávneho kraja". Košický Samosprávny Kraj, http://www.vucke.sk/APIR/sk/Pre_Podnikatelov/Investicne_prostredie/regionalnainovacnastrategia/Documents/RIS.pdf (accessed February 5, 2011).

[64] "Regionálna inovačná stratégia Prešovského samosprávneho kraja". http://www.po-kraj.sk/sk/samosprava/urad/odbor-regionalneho-rozvoja-uzemneho-planu-zivotneho-prostredia/dokumenty-regionalneho-rozvoja/projekt-ris.html (accessed May 5, 2011).

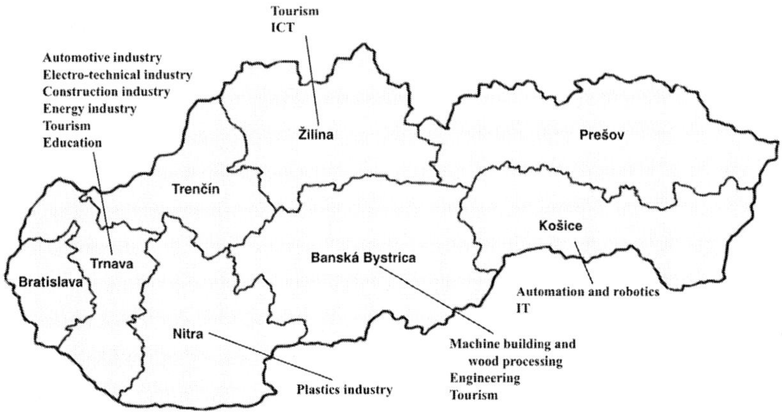

Map 2-4 Division of predominant industries of Slovak clusters/cluster initiatives based on the regions
Source: Own development

Cluster policy in Austria

The type of cluster policy in Austria can be described as a policy of **national frameworks for regional policies**. According to Furre (2007), several ministries and agencies are involved in shaping the cluster policy in the country. There are several strategic national guidelines focused on issues of economic growth, innovation and employment and implicitly on cluster policy also. Nevertheless, the cluster policy appears more apparent and explicit on a regional level, where guidelines for regional development allude to cluster development.

Clusters are actively supported within a technology policy conducive to innovation development. The beginnings of cluster policy development in this country date back to 1995. The Institute for Industry Studies (IWI) and the Austrian Centre for Economic Research (WIFO), following Porter's approach, undertook efforts at identifying clusters in Austria and formulating conclusions for the policy (Lehner et al. 1998, Peneder 1994, Tödtling 2001, Weiss 1994). These two initial studies constituted the basis for the development of further approaches, for example the Hutschenreiter (1994) approach, who tried to identify technology clusters or detailed studies of particular clusters under the TIP Programme (Technology, Information, Policy consulting).

The main impulse consisted of recommendations of a cluster-focused subsidy policy in guidelines for a technology policy in the region of Styria, which consequently took the lead and became a cluster pioneer in the country. Since this time implementation of regional cluster policies has been embraced by all the Austrian regions. Numerous public and private actors, i.e. ministry departments, agencies, banks, public partnerships, non-profit organisations, etc., on a national as well as regional level, have been engaged in the clustering process. Eight national and additional regional cluster programmes have been implemented.

The **National Action Plan for Innovation**, implemented as part of the Austrian Reform Programme for Growth and Employment within the time frame of 2005-2008, highlighted certain aspects concerning the cluster policy in the country. Among others, it emphasised the necessity for the co-ordination of the subsidy policy of the federal states, the co-operation platforms of the regional cluster initiatives, support for R&D co-operation in science-industry and stimulation of international co-operation for the Austrian cluster initiatives.

The main objectives of the cluster policy in Austria have been aimed at fostering regional economic development, supporting co-operation, and innovation and technological development of companies in particular SMEs, thus ensuring their competitiveness, along with supporting the development of centres of excellence.

Cluster support in Austria is focused on all forms of networking co-operation, i.e. clusters, networks, corporate networks, economic complexes, centres of excellence and technological platforms.

Cluster - supporting authorities and institutions

Numerous institutions take responsibility for the implementation of the cluster policy in the country. The main **supporting authorities on a ministerial level** in Austria consist of:

- The Federal Ministry of Economics and Labour (BMWA),
- The Federal Ministry of Transport, Innovation and Technology (BMVIT),
- The Federal Ministry of Science and Research (BMWF), and
- The Federal Ministry of Finance (BMF).

The following bodies play a major role in pursuing cluster policy:

- Austrian Research Promotion Agency (FFG),

- Austria Economic Service (AWS),
- Austrian Federal Economic Chamber (WKO), and
- Austrian Business Agency (ABA).

Apart from these, each of the nine Austrian regions (federal states) has its own **regional policy bodies,** i.e. organisations or agencies, which are involved in cluster issues. Their role in support of cluster concept development is significant as these institutions are focused on supporting and strengthening businesses in particular regions, stimulating innovations and research development, acting as a central contact point (one-stop-shop) for companies or simply promoting the co-operation and formation of cluster networks.

Among them the following can be distinguished (Furre 2007):

- Vienna City: Vienna Business Agency (WWFF), Centre for Innovation and Technology (ZIT) and Vienna Science and Technology Fund (WWTF),
- Lower Austria: Ecoplus (Business Agency of Lower Austria) and Fund for Business and Tourism Development Lower Austria,
- Upper Austria: Upper Technology and Marketing Corporation (TMG), Clusterland Oberösterreich GmbH (Clusterland Upper Austria) and Upper Austria Energy Agency,
- Styria: Styria Business Development Corporation (SFG),
- Burgenland: Burgenland Economic Service (WIBAG),
- Salzburg: Salzburg Innovation Service and Innovation and Technolgy Transfer Salzburg (ITG Salzburg),
- Tyrol: Tyrolean Future Foundation,
- Vorarlberg: Vorarlberg Business Location Agency (WISTO), and
- Carinthia: Carinthia Economic Promotion Fund (KWF) and Centre for Innovation and Technology (Carinthia Development Agency).

Additionally, local authorities play an important role in the initiation and management of clusters.

Generally, clusters occupy an important place in the policy for regional development. The policy makers strive to strengthen already existing links among companies, research institutions and other bodies through facilitating the information flow and creating favourable conditions i.e. law regulations, stimulating an increase in the quality of the labour force, creating demand for public orders, etc. for the development of networking and clusters (Pilarska 2010).

In 2006, the **Clusterland Upper Austria** (further "Clusterland") was established in the region of Upper Austria, becoming an umbrella for six cluster organisations and three cross-sectoral networks. Under the patronage of Clusterland, the **Cluster Academy** is being organised providing an excellent opportunity to learn how to develop a successful cluster initiative, learn from the experience of others in regard to successful cluster organisation and management, as well as serving as a platform for the exchange of information and valuable practical know-how.

Another organisation providing active assistance in the formation of co-operation links and cluster networks also deserves attention, this being **Ecoplus** in Lower Austria. It constitutes "the interface between business and politics, companies and authorities, investors and initiators of regional projects-at both the national and international level"[65]. Since 2001, Ecoplus has established six cluster initiatives in strong and promising industries such as plastics, green building, food, logistics, automotive and mechatronics. It assists in their development and management. Furthermore, the agency's cluster teams:

> support companies in the realisation of their projects and ideas, broker co-operation partners, accompany projects to completion and form the interface to R&D facilities as well as to the public sector[66].

The need to undertake further steps on the Austrian cluster landscape, in the sphere of co-operation between clusters of different regions and co-operation between clusters and universities as well as national cluster initiatives, has been identified by the Austrian Council for Research and Technology Development (RFTE). The Council, in its recommendations to Austrian cluster initiatives, outlined the necessity for a tighter connection between Austrian cluster policies and European cluster policies as well as a need for a stronger focus on national cluster programmes. Additionally, it recommended and supported the establishment of a platform for clusters on a national level (RFTE 2009). In 2008 a **Clusterplattform Österreich** (Cluster Platform Austria or the National Cluster Platform of the Republic of Austria) was consequently founded by the Federal Ministry of Economy, Family and Youth of the Republic of Austria. The platform provides not only information about clusters in Austria, but also information for the clusters about cluster funds as well as regional, national and European cluster policies. The platform comprises

---

[65] http://www.ecoplus.at/en (accessed May 6, 2011).
[66] Ibid.

four working groups active on the following topics: Austrian Cluster Politics; Cluster, Science and Innovation; European Integration of the Austrian Cluster Policies; Cluster and Internationalisation (Clusterplattform Österreich 2011).

An overview of government programmes of support for clusters

Cluster politics and policy in Austria constitute a complex and multi-layered system, which derives from the EU-Programmes Structural Funds, Framework Programme 6 and 7 and the CIP. Numerous programmes exist on a national level deriving from the national strategy framework "strat.at 2007-2013" provided by the promotion agencies FFG and AWS. The national cluster policy is supplemented within the regional cluster policies and programmes in the federal states, where clusters are partly owned by regional initiatives, which provide them with financial support and serves as a platform for co-operation for clusters (Clement 2010). Table 2-1 provides an overview of currently available cluster programmes as well as policies and institutions they derive from.

Amongst others, the **Competence Centres for Excellent Technologies** (COMET) Programme deserves attention. COMET, focused on initiating and supporting long-term co-operation and research in the science-industry sphere as well as on integrating international research expertise was launched for the time horizon 2006-2017. Support within this programme is prioritised according to three categories with special selection criteria: K-projects - with the objective of initiating high-quality research in science-industry co-operation with a medium-term perspective; K1-centres - with the objective of initiating high-quality research with a medium to long-term perspective; and K2-centres - with the objective of bundling the existing national expertise and players in the long-term and co-operation with the world's leading partners (Furre 2007).

Another programme providing several grants is **Co-operation and Networks for Innovation** (COIN), which is supporting clusters within its detailed programme "Co-operation and Networks".

In addition, further programmes with different time horizons were implemented on a national level such as: **Protec-Net plus** (2002-2006) focused on technology transfer and aiming at the establishment of platforms and networks of co-operation in order to stimulate innovation development; **REGplus** (2000-2006) focused on strengthening the existing business incubators and technology centres in the regions; **CIR-CE** (Co-operation and Innovation and Research with Central and Eastern Europe 2005-2008) promoting the establishment and implementation of transnational innovation

Table 2-1 Overview of the Austrian cluster policies and support programmes

The Austrian cluster policies		
International/European	**National**	**Regional**
• EraSME • The Europe INNOVA TM initiative under the Competitiveness and Innovation Programme (CIP) • Central Europe • Programmes under FP7: - Regions of Knowledge - Research for SME Associations - Research for SME's • Small Business Act (SBA) • European Knowledge Intensive Services Innovation Platform (KIS-IP) • Enterprise Europe Network • European Investment Fund (EIF) • CORNET • INTERREG IVC	•Service Initiative: BMWFJ, FFG •AWS-Programmes: - Impulse - Creative Industries - ProTrans •FFG-Programmes: - COMET - Competence Centres for Excellent Technologies - COIN – Co-operation and Networks for Innovation - BRIDGE – Brückenschlag-programm - Innovationsscheck Österreich - Basis Programme - Research Studios Austria - Josef Ressel-Zentren - R&D Programmes linked to certain topics: o TAKE OFF o IV2SPLUS o FIT-IT o BENEFIT o AT:net o Service-IT o NANO o GEN-AU o KIRAS	•**Burgenland**: 1 programme supporting co-operation, •**Upper Austria**: 3 programmes supporting co-operation and networks, •**Lower Austria**: 1 programme supporting co-operation, •**Salzburg**: 1 programme supporting co-operation, •**Tirol:** 4 programmes supporting co-operation and research, •**Vorarlberg**: 1 programme supporting co-operation, •**Vienna**: 2 programmes supporting co-operation and R&D.

Source: www.clusterplattform.at; Federal Ministry of Economy, Family and Youth of the Republic of Austria

co-operation between Austria and partner countries from CEE; **FIT-IT** (since 2001) aimed at fostering development in the field of information technology; and **Go international** (2003-2007) undertaking 32 action programmes supporting the internationalisation of the Austrian economy, elimination of its weaknesses and strengthening its competitiveness.

Furthermore, Austria introduced a special programme named **LISA** (Life Science Austria) focused on support of one particular cluster, Life Science Austria. The programme is not time-limited and also has its equivalent on a regional level, where the Life Science Austria Vienna Region (LISA VR) cluster was established.

The cluster concept has been developed both on the national and regional levels, however, with different intensity and steps undertaken.

Since cluster policy in Austria is implemented on a regional level within a technology policy conducive to innovation development, particular regions indicate a different level of activity in supporting clusters. The region of Upper Austria (the "follower" of the very first efforts of the region of Styria) is particularly active, which, within its systematic innovation strategy, has applied a series of strategic programmes focused on cluster set-up such as "Upper Austria 2000+", professionalising of cluster and network initiatives within "Innovative Upper Austria 2010" and expansion of clusters in order to form professional co-operation platforms within the currently realised "Innovative Upper Austria 2010+". The region targets improvement of competitiveness and its innovative strength, aims at supporting enterprises, particularly SMEs, inter-branch know-how transfer and stimulating development of innovation through co-operation and competence. The region of Upper Austria has begun to be considered Europe's leading region in regard to cluster and networking organisations.

The current cluster landscape in Austria

Clusters in Austria began to develop as early as the middle of the 19th century, when the very first bio-pharmaceutical cluster started to formulate at a Biopharma-consortium in Tirol (ABA 2008). In the 1980s the AOEM (Austrian Original Equipment Manufacturers), a cluster-like consortium in the automobile industry was established. In 1990 a consortium of the woodworking industry called "ProHolz" was started, which has been the basis for wood clusters in several federal states. The modern concept of clusters spread rapidly in the 1990s (Clement 2010). At present, the cluster landscape in Austria is flourishing with more than 50 clusters already established and becoming an important economic factor in Austria (Clement and Welbich-Macek 2007). The Austrian cluster landscape is rich and diverse over several fields of industry. More than 50 clusters encompass around 3,500 member companies in total, which involves up to approximately 420,000 employees and an estimated total turnover of 80 billion Euros (Clement 2010). The Austrian cluster landscape is one of the most developed ones in Europe, which derives from the tradition of business associations that share several similarities with clusters, such as providing information for the members, organisation of training and joint PR-activities, etc. Nevertheless, there are differences with a stronger regional focus on clusters and the fact that innovation activities are more linked to clusters than to traditional business associations (Porter 1998). More than 50 institutionalised and actively operating clusters in Austria are located in nine different fields of industry. Table 2-2 shows the division of clusters based on the industry and the number of clusters active in particular fields. The highest number is visible in the field of renewable energy and environmental technology, where ten clusters were established.

Table 2-2 Cluster industries in Austria
Source: www.clusterplattform.at (accessed May 6, 2011); Federal Ministry of Economy, Family and Youth of the Republic of Austria

Industry	Number of clusters
Automotive, railway, traffic/transport, aerospace	5
Food technology	4
Health, life sciences, wellness	6
Human resources, design, multimedia	8
Information, communication, processes, logistics	6
Materials, packaging	5
Mechatronics, electronics, informatics, sensor technology	5
Renewable energy, environmental technology	10
Timber, furniture, accommodation, building construction	7
Total	**56**

Map 2-5 depicts the division of predominant industries of Austrian clusters/cluster initiatives based on the federal states.

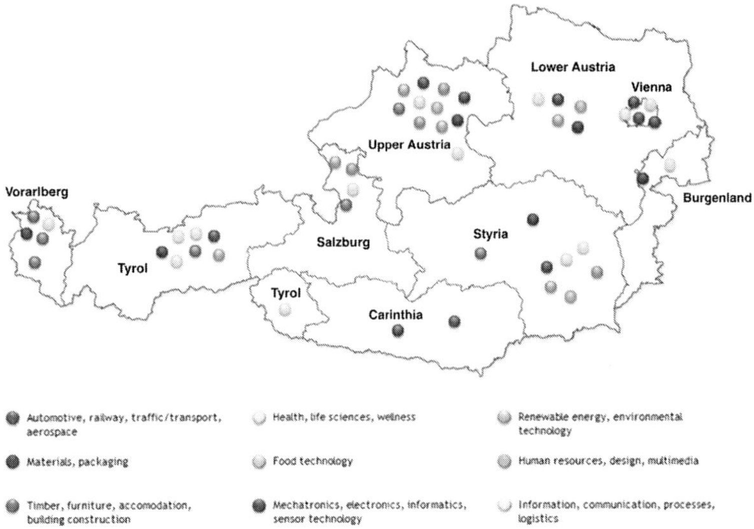

Map 2-5 Division of predominant industries of Austrian clusters/cluster initiatives based on the federal states
Source: Adapted from the official website of the National Cluster Platform of the Republic of Austria: www.clusterplattform.at (accessed May 6, 2011); Federal Ministry of Economy, Family and Youth of the Republic of Austria.

In Austria, clusters typically emerge bottom-up, deriving from industrial needs and developments, not through active involvement of a regional agency or institution. Several regional agencies however, such as ICT Burgenland, Ecoplus, ITC Salzburg, Clusterland, Tyrolean Future Foundation, KWF, SFG, ZIT and WWFF, are in charge of the operations and financing as well as being involved in the ownership of the institutionalised clusters.

As the focus of policies and the number of funding programmes vary in the federal states, the differences in the cluster landscape throughout Austria are visible in the number of established and institutionalised clusters. Having eleven institutionalised clusters, Upper Austria is the most active federal state, followed by Tirol and Styria, which have nine active clusters each (Table 2-3). The least number of clusters can be observed in the federal states of Burgenland and Carinthia.

Table 2-3 Number of clusters according to the federal states of Austria

Region/Federal State	Number of clusters
Burgenland	2
Carinthia	2
Lower Austria	4
Lower Austria/Upper Austria	1
Lower Austria/Upper Austria/ Salzburg	1
Salzburg	4
Styria	9
Tirol	9
Upper Austria	11
Vienna	8
Vorarlberg	5
Total	**56**

Source: www.clusterplattform.at (accessed May 6, 2011); Federal Ministry of Economy, Family and Youth of the Republic of Austria

Table 2-3 also depicts the development of cluster co-operation between regions such as the Plastics Cluster operating successfully across Lower and Upper Austria or the Verein NETZWERK Arbeit Gesellschaft network operating across Lower Austria, Upper Austria and Salzburg.

Comparison of cluster policies in the selected countries

The European Union's desire to make Europe the most competitive and dynamically developing area in the World has resulted in stimulation of innovations and mobilisation for cluster concept development. Therefore, numerous initiatives have been launched in this direction. The EU has based its cluster policy on activities focused on the creation of an environment favourable for cluster development, i.e. ensuring appropriate financial instruments and improvement of co-ordination channels, supporting science-education-industry co-operation, etc. The levels of intensity of the activities undertaken and models of implemented cluster policy in the member countries, however, vary significantly. Table 2-4 presents a comparison of cluster policies in the selected countries.

Table 2-4 Comparison of cluster policies in the selected countries

Country	Type of cluster policy	Main characteristics
Poland	No explicit cluster policy	- 2000-2004 - an adopting period for a cluster policy, - programmes primarily implemented on the EU and national level, - focus on promoting clustering, improvement and development of competitiveness of the existing cluster structure, of co-operative relations, enhancing trans-national co-operation, and stimulating development of a transparent cluster policy, - clusters are supported within different policies, mainly innovation and regional development policies.
Czech Republic	No explicit cluster policy	- 2000-2004 - an adopting period for a cluster policy, - focus on sustainable development of cluster initiatives and a cluster policy in the country, stimulation of clusters' internationalisation, - strong financial support for clusters within national programmes, - clusters are primarily supported on a national level.
Slovak Republic	No explicit cluster policy	- adoption of a cluster policy after 2005, - focus on preparation of strategic materials for development of a cluster policy in the country, - lack or marginal support for clusters.
Austria	National frameworks for regional cluster policies	- beginnings of a cluster policy date back to 1995, - numerous programmes of support on national and regional levels, - strong focus on science-industry co-operation, professionalisation of clusters and network initiatives and stimulation of their internationalisation, - clusters occupy an important place in regional development policy, - clusters are actively supported within technology policy.

Source: Own development

Although Poland does not have a comprehensive national policy based on clusters, attempts at creating favourable conditions on a policy level are being observed. Awareness of the clustering idea began to flourish after joining the EU structure, with the implementation of the first programmes promoting the cluster concept in the country. Since that time, various support mechanisms have been offered within various programmes implemented on a national as well as a regional level. In addition to this, clusters can also use possibilities to apply for subsidies available on the EU level. An analysis of the Polish cluster-based policy, however, indicates a number of deficiencies. The main barriers are connected with the difficult accessibility of the provided sources and their dispersion, as well as a lack of stable, comprehensive guidelines for cluster support in the country. Nevertheless, an endeavour to support clusters and cluster policy development can be seen, as confirmed by the appointed special Working Group on Cluster Policy within the recently conducted PAED project.

In the Czech Republic, the "boom" in cluster concept development began in 2004 with an educational programme on clusters offered by CzechInvest. The OPIE programme explicitly focused on clusters and the subsequently implemented National Cluster Strategy resulted in great popularisation of the cluster phenomenon in the country. The initial enthusiasm, however, disappeared with bureaucracy and organisational problems, which appeared within the implementation of the programme. In addition, the initiated cluster strategy did not meet the raised expectations and has not been continued within the next programming period. However well-launched, the initial steps towards the development of a cluster policy in the country have been neglected. The supporting mechanisms for clusters are currently being implemented primarily within the OPEI programme with an undefined vision for further clusters and cluster policy development in the country.

The Slovak Republic ranks among those countries which have adopted the cluster-based policy the slowest in comparison with the other presented countries. Thus far, the cluster concept has been treated marginally and only in recent government documents and programmes has the topic of clusters been undertaken.

The cluster concept is relatively new in the Central European countries of Poland, Czech and Slovak Republics. Major programmes supporting clusters began to appear in these countries after their accession to the European Union. The current Operational Programmes for the years 2007-2013 held within all three countries underline the importance of clusters as tools for company competitive strategic development.

Austria appears to be significantly ahead in the development of its cluster policy, providing numerous programmes of support both on a national as well as on a regional level. Although the cluster policy and its framework are being shaped on a national level, its implementation appears explicitly on a regional level. Clusters are actively supported within technology policy. They occupy an important place in the regional development policy, where local government plays a significant role in the processes of initiation and management of clusters. Cluster policy in Austria constitutes a complex system with a strong focus on science-industry co-operation, professionalisation of clusters and network initiatives as well as on stimulating internationalisation of Austrian clusters and cluster initiatives.

CHAPTER THREE

CLUSTER ACTIVITIES SUPPORTING CLUSTER STRATEGY

Clustering is based on a team activity, not an individual effort. Cluster actors implement common activities in various areas within, for example, common projects, research, production and services, both with a short-term and long-term perspective, which have a place throughout a cluster's life cycle.

According to Pavelková et al. (2009), those activities which are jointly implemented by a group of cluster members or such activities (services) which are organised for example by the cluster management (cluster organisation) and provided to the members are considered common cluster activities. Common cluster activities should support cluster strategy and should always be carried out in connection with cluster visions and goals, and in regard to established priorities. Cluster activities usually follow a certain path starting with creating a platform for mutual co-operation of companies (networking), followed by common projects implemented in sales, human resources, production and marketing areas and finally projects in the area of R&D.

Cluster life cycle and activities

Each phase of the life cycle of the cluster requires appropriate activities which take place in the cluster. In the phase of cluster formulation, it is important to create networking and partnerships amongst cluster members. Public resources play a significant role in cluster support. This support may encourage collaboration and provide the necessary information for cluster development. Van Klink and De Langen (2001) emphasise that during the cluster development phase; co-operative activities include research and development, the development of co-operative routines and standardisation. During that stage, critical demand stimulates the development of new products and better methods of production. In the expansion phase, clusters seek to expand by entering new markets with the activities including R&D, education and marketing, refining the production process and enlarging the product range. Sharing infrastructure (physical, information or knowledge infrastructure) takes place as well. In the consequent maturation phase, cluster activities include education, research and development and marketing. With the maturing of the cluster, it is vital to place an emphasis on developing specific forms of co-operation and encouraging joint projects amongst companies. At the maturity or decline stage, it is necessary to find new forms of partnerships due to the changing market conditions and potential threats. At this stage it is appropriate to focus on innovation activities. Clusters focus on R&D, education and/or new co-operative routines. In this phase either a decline of the cluster or the development of new co-operation can be observed.

As Pavelková et al. (2009) aptly states, in the initial stages of cluster development, the focus is placed on short-term tactical activities. "Low hanging fruits" is typical jargon in cluster activities, this being a term for fairly easy activities (with low-risk) implemented in the early stages of the development of the cluster (e.g. developing a web-site with interactive links to local firms, visits for cluster stakeholders, identifying stepping-stones through a transparent workshop process). These activities do not require special resources, but do provide quick results whose successful accomplishments builds trust among the cluster members, necessary for the implementation of sophisticated strategic activities.

Each cluster should have an initial portfolio of initiatives prepared which bring benefits to a large number of cluster participants, and provides risk diversification. Not every initiative will bear fruit, therefore, at the beginning it is necessary to focus on initiatives with low risk and quick returns. The cluster can otherwise become extremely vulnerable, when, apart from recurring joint meetings, it will not generate any specific outputs over many months. Representatives from the private sector (esp. SMEs) may lose interest in the initiative.

Correct project identification is paramount to the initial success and for unifying the steering (management) group. Following the completion of the initial less intensive activities, the cluster's focus shifts towards long-term and substantive projects. In this phase, an update of the initial strategic decisions can also often be observed.

Certain activities should be implemented, however, regardless of the stage of the life cycle, with merely changes in their intensity and importance.

Focus on cluster activities

On the basis of a great deal of research conducted[1], cluster activities can be divided into the following areas:

1. Networking;
2. Human resources;
3. Research and innovations;
4. Business co-operation and promotion;
5. Financing investment plans;
6. Governmental/political area (lobbying);
7. Supporting activities.

Networking activities involve the formal and informal gathering of information within the cluster. Sharing information can be done through meetings of cluster members' representatives, information platforms such as websites, common newsletters and magazines etc., joint seminars, workshops and short courses for firms and organisations within clusters amongst others. Also contacts with actors outside the cluster, for example suppliers, customers, providers of funds, research and education institutions and other entities, have to be taken into consideration. Networking activities stimulate the exchange of knowledge, experience and good practices among cluster participants, and enable them to learn from one another and to build trust and communication linkages inside and outside the cluster.

Within the area of human resources, clusters focus on increasing the skills and qualifications of their members. These activities include education and training (the exchange of knowledge and experience), organisation of seminars, workshops and conferences, co-operation with educational institutions, etc.

Research and innovation activities, facilitating improvements in production processes and enhancing technology development, constitute an inseparable element of cluster activity. They are focused on common R&D, co-operation with research institutions, implementation of innovative projects (regarding production processes, products etc.), support of spin-off creation and incubator services.

---

[1] Sölvell, Lindqvist and Ketels (2003), Innovating Regions in Europe. "IRE subgroup on Regional clustering and networking as innovation drivers: Cluster Management (Learning module 5)", "Cluster Management Guide". CLOE (2006), Pavelková et al. (2009).

Business co-operation and promotion activities also play a crucial role in cluster development. Regarding the business co-operation area, cluster members focus on joint purchasing, sharing production, and logistics management. In the area of promotion, they focus on marketing research of trends and markets; join participation in trade fairs, joint catalogues of products and services, logos, trade names, and joint advertisements.

In the area of financing investment plans, clusters focus their activities on joint investments in the infrastructure, obtaining appropriate and sufficient financial resources, mediacy in obtaining support services (banking services, insurance, legal, accounting and tax consultancy, etc.).

A further important area of co-operation involves activities connected with lobbying and influencing the policy sphere, for example lobbying in favour of the infrastructure, for more favourable legal regulations or for subsidies.

Clusters can be also accompanied by support activities such as helping in the preparation and management of projects, benchmarking and other support services.

Due to the cluster characteristic, it should be emphasised that the effects of joint activities are often visible on the long-term time horizon. Since the results of collaboration may appear after a period of several years, in the meantime, the risk of losing interest in common initiatives may appear from the side of participants who may be impatient and tired of waiting for them. It is therefore crucial to build strong links between the participants with a developed culture of trust and co-operation, and ensures internal and external stability. This can be obtained by a portfolio of cluster activities which brings benefits to a large number of cluster participants and provides risk diversification. This is why the implementation of simpler and easily reachable, low-risk activities providing quick-wins is being observed at the beginning of cluster activity. These, thus, encourage the activity and confidence of members for the subsequent implementation of more strategic activities.

Examples of cluster activities from Poland, the Czech Republic, the Slovak Republic and Austria

Clusters realise their goals through the joint activities of their members (cluster actors) which are set according to the industry field, their specialisation and resources of the region in which these clusters are located. Even though these activities may be specific to each cluster, they appear as common in principle.

The conducted research focused on the areas of cluster activities specified above with the attention to individual activities distinguished in particular groups.

A total of 159 clusters from the entire world participated in the research; of that 88% were European clusters. The research took place gradually in the year 2007 (with the participation of 53 clusters, or 33%), and then during the years 2010 and 2011 (with the participation of 106 clusters, or 67%). Clusters participating in the research were from different countries and consequently from economies with different levels of development, where 30% of the respondents were from countries with "more advanced economies" and 70% from countries with "developing economies".

This publication concentrates only on several selected countries, specifically Poland, Czech Republic, Slovak Republic and Austria. Therefore in the following text only clusters (respondents) from these countries will be analysed; their total number is 116. The composition of the sample is shown in Figure 3-1.

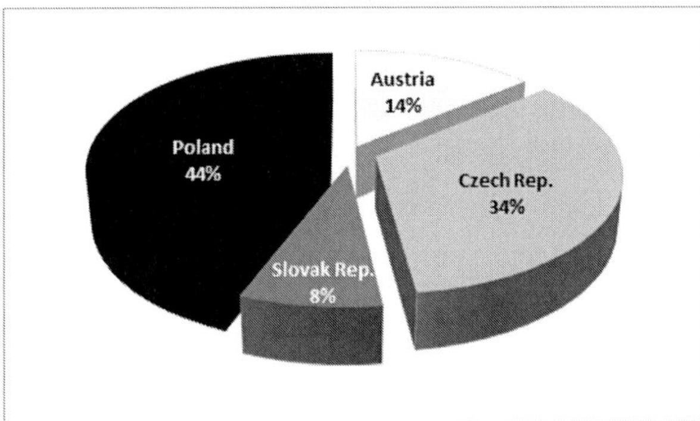

Fig. 3-1 Division of respondents according to the country of origin
Source: Own development

The highest proportion of clusters participating in the research was from Poland[2] and a significant proportion from the Czech Republic. The very low proportion of clusters from the Slovak Republic is related primarily to the shortest experience among the selected countries,

[2] It should be pointed out that according to the latest nationwide research conducted by PARP (2012), over 250 clusters/cluster initiatives were identified in Poland.

considering the cluster concept development and thus the low number of already existing clusters or potential clusters and cluster initiatives (more in the section: Cluster policy in the Slovak Republic).

Concerning the external environment (macroeconomic environment) of clusters, only the clusters from Austria operate in the environment of advanced economy. The remaining three countries can be characterised as "developing economies". In all cases, they are post-communist countries. Therefore, it is correct to assume that the conditions for the formation and development of clusters in these countries will be different. Also the substantial characteristic of the given sample of the clusters is their age (Table 3-1). When excluding the natural (porterian) clusters, the age of clusters (particularly in the Czech Republic, Poland and Slovak Republic) unfolds depending on the history of the introduction of cluster policies in the individual countries. It should be pointed out that this research focused on organised clusters. The absolute majority of clusters (86%) from the sample can be included in the category of young (embryonic) clusters. These are the clusters that are active for a maximum of five years. It is necessary to emphasise that in this category can be included almost all clusters from the Czech Republic, Poland and the Slovak Republic. Only in the case of Austria is there a greater proportion of more established clusters; this is certainly tied to the earlier and more advanced development of both the cluster concept and cluster policy in this country (more in section: Cluster policy in Austria).

Table 3-1 Division of embryonic and established clusters according to the country of origin

	Embryonic clusters	**Established clusters**	**Total**
Poland	48	3	51
Czech Republic	39	1	40
Slovak Republic	8	1	9
Austria	5	11	16
Total	100	16	**116**

Source: Own development

While analysing the level of implementation of individual areas of activities, from the results it is obvious that the most usual activities of clusters are in the **area of networking**, which is true for all studied countries. These activities are implemented by about three quarters of all clusters (while in Austria this proportion is somewhat greater but lower in the Slovak Republic and Poland), however the percentage of clusters, which are already implementing or plan to implement them is 90% (Figure 3-2). This is due to the fact that the process of clustering is not a closed process and the continuous drawing in of new subjects is important for the progress in co-operation. Another result is that there is a growing supply of potential leaders with energy, vision, local contacts and the ability to exploit other resources. For highly efficient clusters the existence of formal and informal social networks is indispensable. The dialogue and mutual trust are based on these ties. However, the process of building trust is necessary for effective co-operation and takes a long time. Therefore, developing ties between the members of a cluster is the first step for creating the necessary trust and co-operation. Often it happens that during the first meeting of cluster members some of them meet the others for the first time. At the same time one of the barriers for the growth of the cluster may be a lack of social networks.

In the next order of activities, it is already possible to note differences between individual countries; a significant difference exists in the Slovak Republic. The second most common activity in all countries except the Slovak Republic are activities in the area of R&D. It is necessary to remark, however, that these activities are implemented to a very high degree by the Austrian clusters (88%). In the case of Czech clusters this proportion is 65%, in the Polish clusters 52% and in the Slovak clusters only 38%. It is evident that these are activities dependent on specific conditions of clusters (they are implemented more frequently by more experienced clusters which attained a certain level of maturity, that is, clusters in an advanced economic environment). It is also necessary to add that in both the Czech Republic and Poland there is a considerable portion of clusters that are planning to put these activities into practice in the future. This can be related to the fact that both Polish and Czech cluster supporting programmes amongst others, aim at boosting technological co-operation, enhancing joint R&D and innovation projects among individual clusters.

The third most often carried out activities are those in the area of human resources. Here again is evident a pronounced difference between Austrian clusters, where these activities are very frequent (83%) and the remaining countries. In the Czech Republic and Poland more than half of the clusters put them into practice, in the Slovak Republic 44% of all clusters. Equally as in the previous case, when comparing the proportion of clusters that are carrying out and planning these activities, this proportion

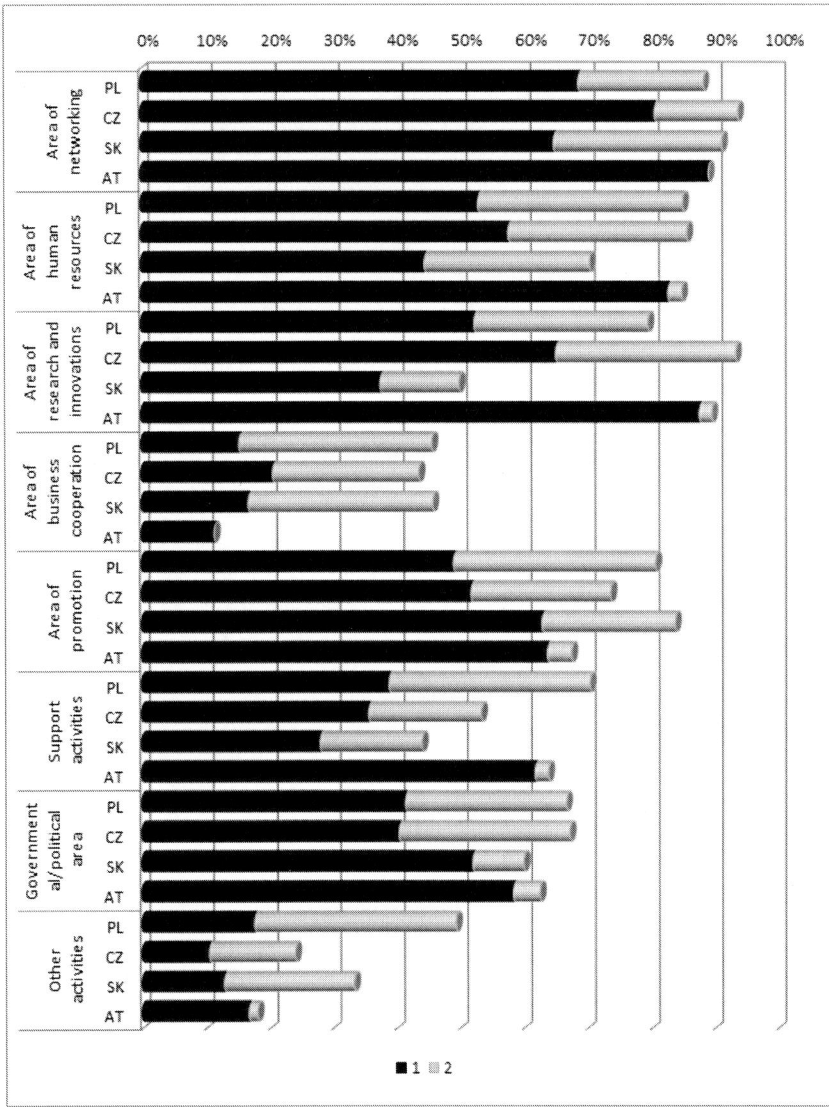

Note: 1 = clusters currently offer the activity; 2 = clusters currently do not offer the activity, but plan on doing so in few years

Fig. 3-2 Main areas of clusters activities according to the country of origin
Source: Own development

is entirely on the same level as in Austria, namely 85% (with the exception of the Slovak Republic at 70%). Clusters realise the importance of social capital and the presence of a skilled and dedicated workforce for the development of their network; therefore, they aim to invest in the creation of high-quality interpersonal links among their members.

Activities in the area of advertisement are in fourth place in frequency. Here, it is interesting that these activities are carried out by a large portion of Slovak clusters (in them promotion is much more frequent than human resources and R&D) and is comparable with the activity of Austrian clusters (64%). This may be connected, on the one hand, with the very young age of Slovak clusters (where promotional activities belong to fairly easy activities with quick results), and on the other hand, with the area of industry they operate in (where significant parts of clusters operate in services and the tourism sector). In Czech and Polish clusters this proportion is approximately 50%. If adding the clusters which are planning this activity, the proportion becomes about 80%. It is interesting that this activity is not of particular interest (68%), what harmonises with the status of their development (maturity).

Next in the order of frequency are activities in the area of lobbying; they are implemented only by 40% of Czech and Polish clusters and by one-half of the Slovak clusters; in Austria the percentage is somewhat higher (58%). It is interesting to note that Slovak clusters engage in these activities more frequently than in HR and R&D. Especially lobbying for subsidies and more favourable legal regulations are very popular among Slovak clusters, which through these activities hope to increase the awareness of authorities about the cluster concept and change their mentality and marginal treatment of cluster issues in the country. The proportion of clusters that engage in these activities or planning to in the future is stable in all countries under study, namely two thirds of the total. The next area of activities includes supporting activities for clusters. In this category Austria eminently dominates with 62%; in the remaining countries the proportions are around one-third of the clusters. It is interesting though that these activities are planned to a large extent in the future.

If we completely sum up the tendencies in the implementation of the above-mentioned activities, it is obvious that the widest spectrum of activities is implemented by Austrian clusters. In their case, the proportion of clusters that put activity into practice is higher than in the other countries. This measure approaches the proportion of the clusters, after the addition of those that are only planning to implement the activities. From that it follows that the clusters are already in a more advanced stage of the implementation of activities compared to other countries.

Additional activities that were studied are put into practice by only a small portion of clusters. Surprisingly, these activities are in the area of business co-operation and additional activities, among those included are such as joint purchasing, shared production and logistic management, as well as joint investment in infrastructure, obtaining financial resources (for example, from a bank, venture capital), spin-offs creation, incubator services, and the electronic marketplace. These activities are carried out on average by only 15%, invariably in all countries. Another interesting finding is that Austrian clusters, which are more experienced than those in the other countries, are not interested in implementing these activities, not even in the future. Based on these results, it can be inferred that obviously these activities are not important for the co-operation of subjects in a cluster and therefore are not being considered. Only less developed clusters with less experience are planning to develop them.

The results from Slovak clusters, which were commented on earlier, are also very specific. Slovak clusters are favouring lobbying and advertisement; this is connected with the different development in that country. It should be added that the Polish clusters generally implement a lower proportion of the most frequent activities (networking, human resources, R&D) than the Austrian and also Czech clusters, in which case the order of the areas of activities is identical.

The following sections depict detailed analysis of clusters activities according to the country of their origin. In the assessment of activities the first number in the parentheses always shows the proportion of clusters that carry out the activity, the second number the total of clusters that are already implementing the activity and plan to do so in the future.

Detailed analysis of activities of Polish clusters

The order of activities of Polish clusters is very similar to those of Austrian and Czech clusters, only with the difference that these activities are implemented to a lesser extent than in these two countries. Figure 3-3 depicts cluster activities according to their areas. The most common activities are in the area of networking (69%; 88%), followed by activities in human resources (53%; 85%), activities in research and innovations (52%; 79%) and activities in the area of advertisement (49%; 81%). Somewhat less frequently used are activities in lobbying (41%; 67%) and supporting activities (39%; 70%). Similarly, as in the other countries, activities in the realm of business co-operation are not favoured (15%; 46%).

In the case of Polish clusters, as in the Czech clusters, it is obvious that the majority of the activities are to a large extent only planned and are only implemented by the part of the clusters that are interested in these activities.

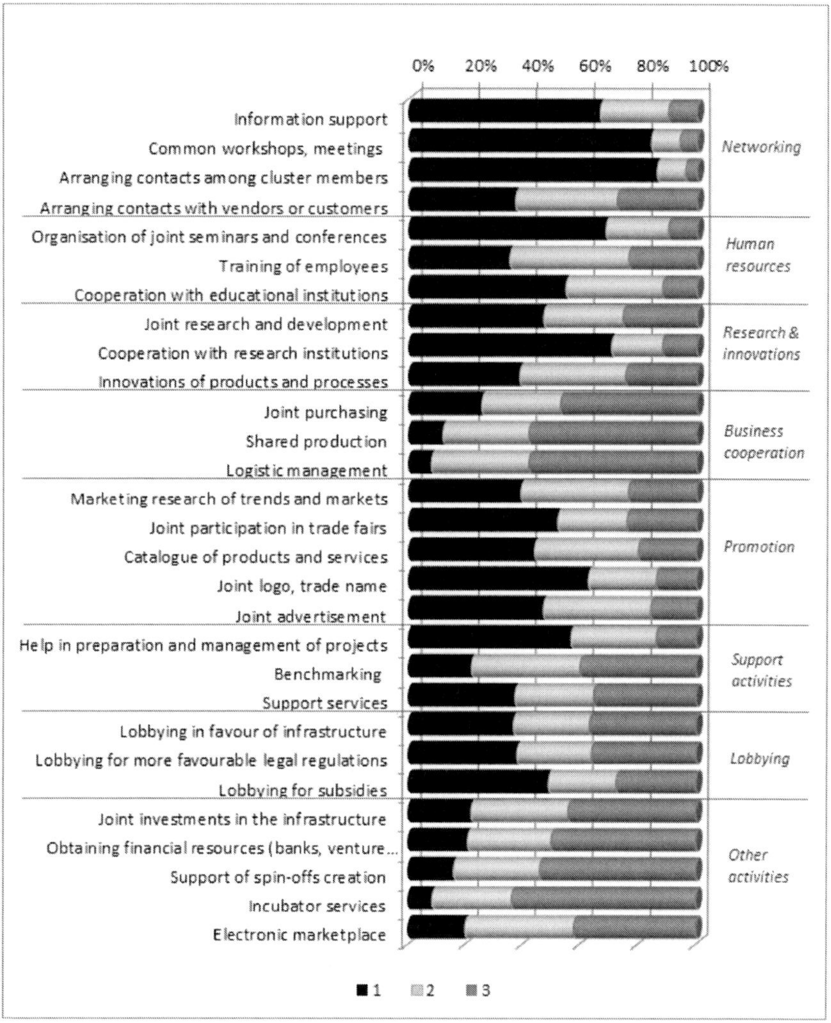

Note: 1 = clusters currently offer the activity; 2 = clusters currently do not offer the activity, but plan doing so in few years, 3 = clusters do not offer the activity

Fig. 3-3 Activities of Polish clusters (in the sample) according to their areas
Source: Own development

In Poland, the area of networking in the future will be as common as it is currently in Austria and in the Czech Republic, however, for the present, a comparatively smaller proportion of clusters carry out this activity. Approximately 85% of clusters attend to arranging contacts among cluster members, as well as common workshops and meetings. In comparison to other countries only two-thirds of clusters provide information support and only one-third are arranging contacts with vendors or customers. Clusters are often in the phase of trust building and overcoming the distance among partners and changing their mentality.

In the area of human resources in Polish clusters prevails the organisation of joint seminars and conferences (69%; 90%). More than one-half of the clusters already co-operate with educational institutions. Again, it has to be emphasised that a substantial proportion of Polish clusters are planning to implement these activities in the future and this is also the case with the training of employees. Investments in human resources and development of co-operative relations between entrepreneurial and educational spheres grow in its importance among Polish clusters.

For Polish clusters, the state of the implementation of activities in the area of research and innovations is specific. Even when talking about the proportion of clusters that plan to implement these activities in the future, this proportion is about 10% lower than it is in Austria and then in particular lower than in the Czech clusters. At this time, Polish clusters chiefly co-operate with research institutions (71%). Joint research and development (47%) and innovations of products and processes (39%) are now, for the most part, only in the planning stage. This is the result of the slow process of learning how to co-operate together and bringing along mutual benefits. Still a low level of awareness regarding the benefits from R&D co-operation among business sphere and research institutions, for example, business or technological parks, and innovation centres, as well as misconception and fear of abuse or exploitation of the results of this co-operation for their own purposes by one of the sides is being observed.

In Poland, activities in the area of promotion are also very popular. In the future they should be implemented in approximately 80% of all clusters. Presently, Polish clusters very often utilise joint logos, and trade names (63%), but somewhat less frequently joint participation in trade fairs (52%). Besides, less than one-half of the clusters are actively engaged in other activities in the field of advertisement.

Activities in the governmental/political area in Poland are currently approximately equally common; lobbying for subsidies slightly outweighs them at this time. The importance of the implementation of these activities is, in both Poland and the Czech Republic, in almost the same condition.

Equally, as in the case of Austria and the Czech Republic, help in preparation and management of projects dominates in the area of support activities. However, it is interesting that in contrast to these two countries a great proportion of Polish clusters (two-thirds) also plan activities such as benchmarking and support services. Benchmarking especially is growing in popularity. Its comparative analysis provides the opportunity to confront clusters with their competitors and to learn from good practices already realised by the others.

Regarding the least carried out activities in the area of business co-operation, Polish clusters equally with Czech clusters concentrate mostly on joint purchasing (a quarter of all clusters). It is equally necessary to point out that in comparison with the Austrian clusters, Polish clusters are planning to implement all these business activities in the future, namely to a substantial degree (approximately one-half of the clusters in all researched activities).

In the case of Polish clusters, it is important to comment that they plan, particularly in the future, to be actively engaged in some untraditional activities, such as the electronic marketplace (20%; 58%), joint investments in the infrastructure (22%; 56%), obtaining financial resources, for example from banks, venture capital (21%; 50%), and support of the creation of spin-offs (16%; 46%).

Detailed analysis of activities of Czech clusters

As was mentioned earlier, for the Czech clusters the absolutely most common activities are in the area of networking (81%; 94%), which can be seen in Figure 3-4. Further are the activities in research and innovations (65%; 93%), in the area of human resources (58%; 86%) and in the area of advertisement (52%; 74%); less often are lobbying (40%; 67%) and supporting activities (36%; 53%). The activities of business co-operation are put into practice by only a low proportion of clusters (21%; 44%). It can be noted that in the Czech Republic there is a major proportion of clusters which as yet are planning the implementation of the large part of their activities in the future.

In the area of networking, Czech clusters similar to Austrian clusters, very often carry out information support, common workshops, meetings and arranging contacts among cluster members. Only one-half of the clusters are arranging contacts with vendors or customers, but in comparison with Austria, this activity is also chiefly planned for the future.

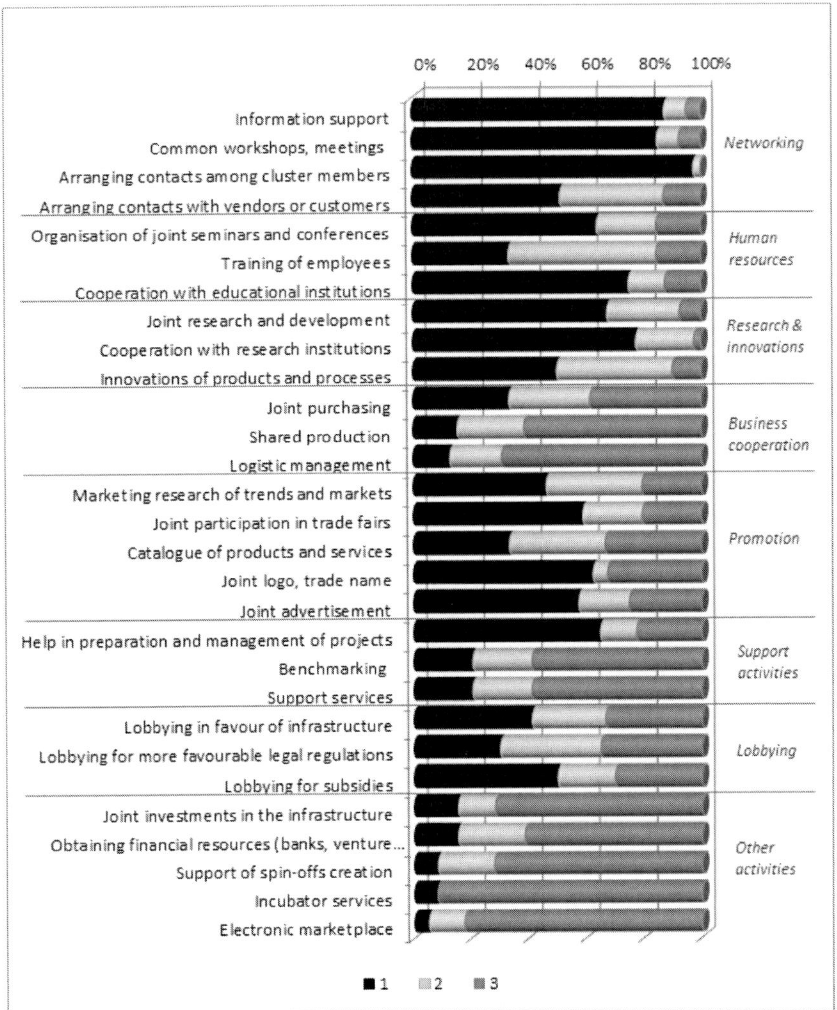

Note: 1 = clusters currently offer the activity; 2 = clusters currently do not offer the activity, but plan doing so in few years, 3 = clusters do not offer the activity

Fig. 3-4 Activities of Czech clusters (in the sample) according to their areas
Source: Own development

Activities in the area of research and innovations represent, according to the Czech clusters, absolutely the most preferred activities, which will be put into practice by almost all clusters. At this time, they are carried out by approximately two-thirds of the clusters; most of that in co-operation with research institutions and in joint research and development. This half of the clusters is actively engaged in innovations of products and processes. The development of co-operation links amongst research and business spheres constitutes a way to improve the conditions for enterprise and innovation, and of gaining the competitive advantage by the entrepreneurs. Also, programmes supporting cluster development in the country are focused on activities leading to innovative improvements and the usage of new technologies, highlighting their importance.

When focusing on the area of human resources, three-quarters of Czech clusters are already involved in co-operation with an educational institution and two-thirds in the organisation of joint seminars and conferences. Only one-third of the clusters are actively engaged in the training of employees, but this activity has a great potential in the future, at which point 85% of Czech clusters should put it into practice. Clusters realise that the importance of strengthening their competitiveness lies in the quality of not only products and production processes, but first and foremost in human resources who are steering the entire mechanism. Therefore, rapidly growing attention is given to different forms of the investment into the development of human resources.

Activities in the area of promotion are also relatively popular among Czech clusters and in the future will be implemented by about three-quarters of clusters. It is interesting to mention that in comparison with other activities in this area, Czech clusters for the time being utilise a mutual catalogue of products and services. Perhaps more attention is paid to the development of the co-operation links within the cluster itself, rather than to the creation of a common product or service.

In the governmental/political area Czech clusters prefer at this time, as do clusters in the other countries, lobbying for subsidies (half of the clusters). A smaller proportion of the clusters are lobbying for more favourable legal regulations and lobbying in favour of infrastructure, but it is interesting that the importance of these activities will almost equalise in time and very likely about two-thirds of clusters will be engaged in them.

In the area of support activities, Czech clusters like clusters in Austria and the other countries, attend mainly to help in the preparation and management of projects (65%; 78%). Other activities in this area are by far not so important.

As for the least implemented activities in the area of business co-operation, Czech clusters count on joint purchasing (one-third of clusters). In comparison with Austrian clusters it is necessary to remark that Czech clusters are planning these business activities in the future at about twice the scale. Clearly, clusters see the advantages in economies of scale and better fluency in production processes.

Detailed analysis of activities of Slovak clusters

The process of the implementation of activities in the Slovak clusters shows entirely different characteristics when compared to other countries in this study. Although Slovak clusters favour networking (65%; 91%), their preferences for other areas of activities are fundamentally different (Figure 3-5). In the Slovak Republic, activities in the area of advertisement (63%; 84%) and lobbying (52%; 60%) significantly dominate. These are followed by activities in the area of human resources (44%; 70%) and activities in the field of research and innovation (38%; 50%). Slovak clusters largely do not implement support activities (28%; 44%) and business co-operation (17%; 46%). In the same way as in Polish and to a degree in Czech clusters, it is apparent that the majority of activities in Slovak clusters will be implemented only in the future. The character of the activities already implemented by clusters in the Slovak Republic reflects the very basic phase of the development of a cluster-based policy in the country. It points out the need for the popularisation of a cluster concept itself and the awareness building about clustering in all spheres of the Triple Helix.

Regarding the activities in the area of networking, in the case of Slovak clusters, it can be stated that they are and will be put into practice by a comparable proportion of clusters as they are in the other countries in this research. Nearly 90% of clusters are arranging contacts among cluster members and almost 80% of them provide information support. In comparison to other countries, only about one-half of the clusters organise common workshops and meetings.

As is already pointed out several times, Slovak clusters currently markedly prefer the area of promotion over other activities. More than two-thirds of clusters utilise joint logos, trade names, attend to joint participation in trade fairs, joint advertisement and marketing research of trends and markets. The proportion of clusters which carry out these activities surpasses the proportion of clusters in the Czech Republic and Poland.

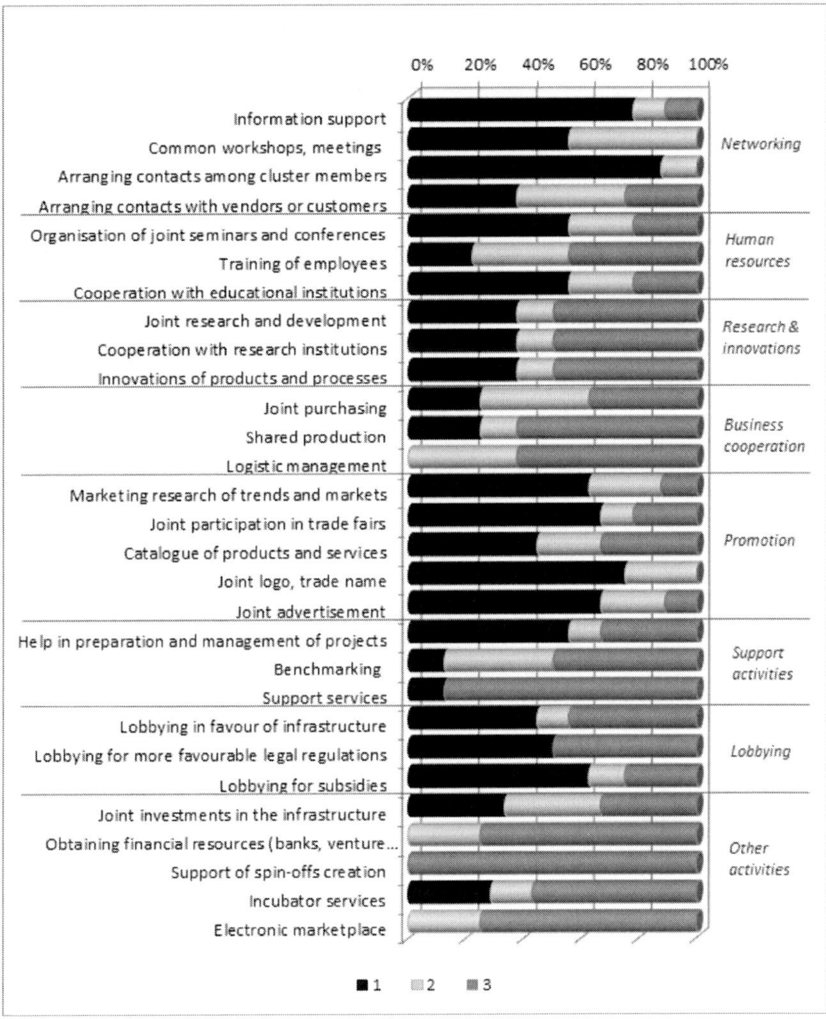

Note: 1 = clusters currently offer the activity; 2 = clusters currently do not offer the activity, but plan doing so in few years, 3 = clusters do not offer the activity

Fig. 3-5 Activities of Slovak clusters (in the sample) according to their areas
Source: Own development

At the present time, Slovak clusters also concentrate intensively on lobbying, particularly on lobbying for subsidies. This is connected with a lack of financial support for cluster development in the country and with a great need for the development and implementation of systemic support instruments for clusters and cluster initiatives. In the future the importance of these activities should be approximately the same as is the case in the other countries.

As for the area of human resources, more than one-half of the clusters organise joint seminars and conferences as well as co-operation with educational institutions; only one-fifth of the clusters engage in the training of employees. These proportions are significantly lower than in the other countries and will not even in future attain the values existing in the other countries. Low awareness and lack of understanding of the cluster concept on the part of the authorities and thus lack of explicitly oriented cluster policy and legislation regarding the establishment and activities of clusters, forces clusters to focus on different priorities at first, simultaneously leaving the development of human capital to future plans.

An even more pronounced difference in comparison with the other countries is observed in the area of research and innovations. All these activities are taken place in only less than 40% of clusters and in the future this should be put into practice by one-half of the clusters. This situation is substantially different from clusters in the other countries, where these activities are absolutely of cardinal importance. It is again the result of a marginal treatment of the cluster concept in the Slovak Republic hampering the development of clusters in general. Low interest in this type of activities may also be connected with the character of existing clusters (where, for example, the tourist sector plays a significant role).

As with clusters in the other countries, Slovak clusters concentrate in the area of support activities mainly with the help in preparation and management of projects. Additional portions of Slovak clusters plan to implement this activity but not to the same extent existing in the other countries.

In the area of business co-operation Slovak clusters intend to devote efforts primarily to joint purchasing (almost two-thirds of all clusters). Assessment of these activities is very similar as it is in the Czech Republic and Poland. All these countries put a greater stress on these activities than is the case of advanced Austrian clusters.

In the case of the Slovak Republic, the high interest in non-traditional activities such as, for example joint investments in the infrastructure (33%; 67%) and incubator services (29%; 43%), is rather surprising.

Detailed analysis of activities of Austrian clusters

In Austria (Figure 3-6) as in the remaining countries, the most common activities are in the area of networking (89%; 89%), but, in contrast to them, are also activities in the areas of research and innovations (88%; 90%) and activities in the area of human resources (83%; 85%). The less frequently carried out activities are those in advertising (64%; 68%), support services (62%; 64%) and lobbying (58%; 63%). Nevertheless, it has to be highlighted that all of them are implemented by more than half of the analysed clusters. Equally, as in the case of the other countries, the activities in the area of business co-operation are put into practice by only a very low proportion of clusters (11%; 11%), and their implementation is not planned even in the future. In the case of Austria, it is evident that the difference between the portion of clusters that are implementing or planning the implementation of given activities is very low, which means that these clusters are already in the stage of development, where they put into practice only such activities that are essential for their progress.

As for the area of networking, all Austrian clusters facilitate information support, common workshops, meetings and arranging contacts amongst cluster members. Only one-half of the clusters are engaged in arranging contacts with vendors or customers.

In the area of research and innovations, the emphasis is primarily on co-operation with research institutions, which holds true with regard to all clusters. A very high percentage of clusters (88%) is engaged in innovations of products and processes themselves and three-quarters of clusters indicated that they carry out joint research and development. It is the result of a culture of co-operation in the field of R&D which is highly ingrained among Austrian enterprises.

In the area of human resources, an absolute majority of clusters co-operate with educational institutions and organise joint seminars and conferences. Three-quarters of clusters are engaged in the training of employees.

As for the area of advertising, there are two markedly dominating activities, namely marketing research of trends and markets and joint participation in trade fairs (both more than 80%). Approximately one-half of the clusters then utilise the catalogue of products and services and joint advertisement, somewhat fewer clusters have joint logos or trade names. It is interesting that clusters from the other countries depend on the use of joint logos or trade names much more than is the case in Austria. In certain cases their usage is considered less convenient especially for the individual

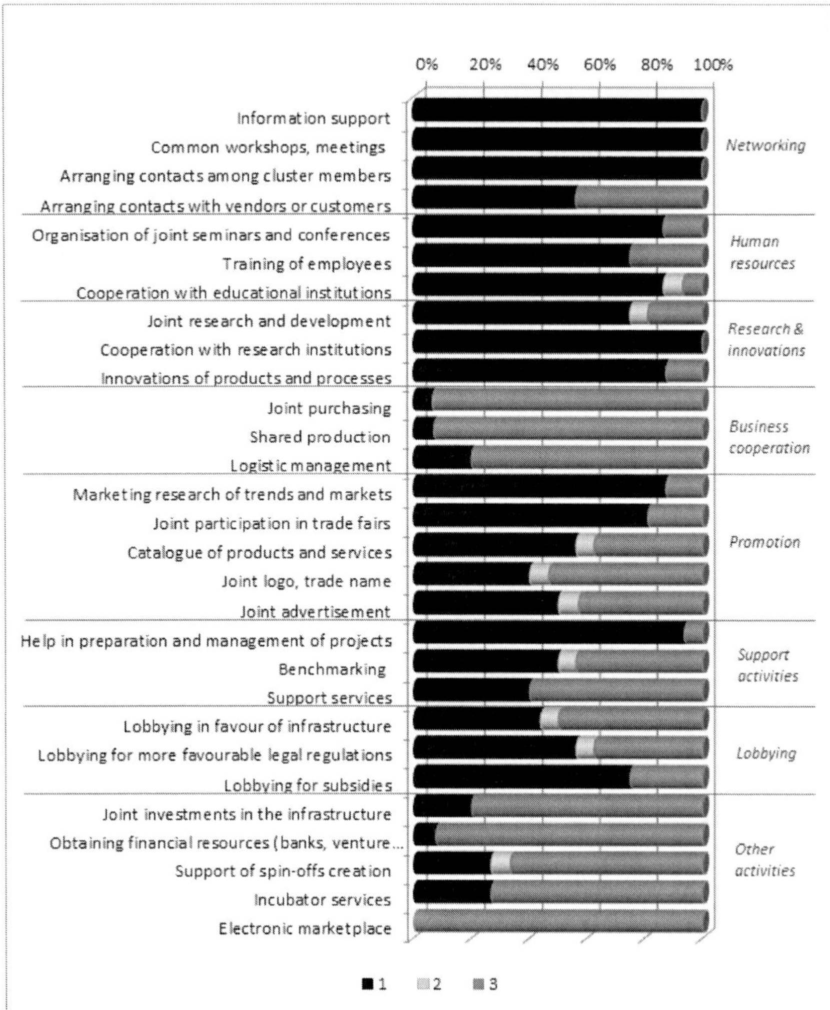

Note: 1 = clusters currently offer the activity; 2 = clusters currently do not offer the activity, but plan doing so in few years, 3 = clusters do not offer the activity

Fig. 3-6 Activities of Austrian clusters (in the sample) according to their areas
Source: Own development

cluster actors whose own trade names are more well-known and bring better feedback in the business sphere than the cluster (which, for example, has younger roots of its existence). This may be one of the reasons why not so much importance is being attached to the joint logo or trade name of a cluster in Austria.

Regarding the support activities, the absolute majority of clusters use help in preparation and management of projects (94%). It is among one of the most common activities of Austrian clusters and which the clusters from the other countries do not carry out on such a large scale. Other activities such as benchmarking and the implementation of support services are not carried out to a high extent (maximum half of the clusters).

In the governmental/political area, Austrian clusters, as with the clusters in the other countries, are favouring lobbying for subsidies (three-quarters of clusters). Approximately half of the clusters are lobbying for more favourable legal regulations and in favour of infrastructure.

It is necessary to mention here that the activities in the area of business operations are implemented only by a negligible proportion of Austrian clusters and if they are, then they favour logistics (20% of the clusters).

It is also necessary to add that some Austrian clusters also carry out other activities, which in the other countries is very rare. These include support of the creation of spin-offs or incubator services (both 27%).

Examples of cluster best practices from Poland, the Czech Republic, the Slovak Republic and Austria

This section depicts a group of clusters characterised by the implementation of interesting solutions in different areas of joint cluster activities. These may serve as good practice examples valuable for other clusters' development, especially for start-up initiatives emerging as well as developing clusters. The presented clusters were selected on the basis of analysis conducted and the evaluation of clusters, which had taken an active part in the authors' research. The examples are shown in the form of a short case study with the focus on best practice solutions in a specified area of activity, and they are the result of close co-operation with cluster managers/representatives (i.e. structured face-to-face interviews as well as close co-operation with the use of modern communication technologies).

Alternatywny Klaster Informatyczny/Alternatywny.klaster.info - PL, Masovian Voivodeship

Alternatywny.klaster.info was founded in Warsaw in 2007 by five companies operating in the IT market or by providing IT services. With its mission of increasing the competitiveness of Polish companies in the sector, currently the cluster associates 27 members from the Masovian, Lublin, Lesser Poland, Subcarpathian and Łódź Voivodeships. Its co-ordinator (the company Sekwencja Sp. z.o.o) provides various services such as support in project management (from searching for suitable projects, help in their preparation and management), IT services (offering lower prices), co-ordination of international co-operation activities as well as entering into new foreign markets, or consultancy in regard to export activity plans.

The development of the joint market information resources is also amongst the aforementioned. Thanks to this activity, cutting edge IT/ICT sector analysis is being provided to member companies. Expert analyses of **marketing research of trends and markets** in the sector are regularly purchased by the cluster co-ordinator who then sends them to the cluster members thus ensuring their access to current information in the sector. In addition, member companies can use the opportunity of on-line access to the updated database of market data and obtain current information about technical and technological innovations (in the area of equipment, software, the telecommunication sector, electronic equipment, etc.), and thus benefit from limiting costs of purchasing services individually.

Stowarzyszenie Grupy Przedsiębiorców Przemysłu Lotniczego Dolina Lotnicza (Aviation Valley, Aviation Valley Association) - PL, Subcarpathian Voivodeship

Aviation Valley is located in south-east Poland, known for its developed aviation industry and pilot training centres. The region is characterised by 100 years of aviation history and 80 years of aircraft industry history, where a large concentration of aerospace companies, research centres, and developed educational and training facilities are located. Most companies involved in Aviation Valley are actually located in the Subcarpathia region, with its heart in the region's capital - Rzeszów. In order to ensure the rapid development of the aviation industry in the region, Aviation Valley, an industry and engineering cluster was established thanks to the initiative of leading companies in the industry in 2003. At the beginning of its development, strong financial support from

Pratt & Whitney, a world leader in the design, manufacture and service of aircraft engines, rocket propulsion systems and industrial gas turbines was received. Its main objective is to transform southeast Poland into one of Europe's leading aerospace regions able to provide a variety of products and services for the most demanding clients. At present, the cluster represents over 90 aerospace companies with further interest shown in entering its structure. Aviation Valley is the first and largest aerospace cluster in Central and Eastern Europe.

Apart from its exceptional production, Aviation Valley provides outstanding education as well. With great attention to the importance of **education in its industry field**, the cluster has stimulated the establishment of an appropriate level of education, starting from elementary education up to university level. The Tech University for Children was consequently established in 2009 and focused on physics, chemistry, mechanics as well as astronomy. Additionally, from the Aviation Valley initiative, a specialised programme to increase the availability of numerically controlled machine tool operators and other professionals associated with the aviation industry was designed in 2005. As a result the Operator Training Centre "CEKSO" was established bringing together technical schools at the secondary-level in the Subcarpathia region. The cluster is also a patron of the Practical Training Centres in 12 towns in the region. In addition, the Rzeszów Technical University is one of the Aviation Valley members. The University has a specialised Construction and Aviation Department training students not only to become aircraft construction and furnishing designers but also offering the opportunity to obtain a pilot's license for professional civil aviation (civil pilot license - CPL) in accordance with the international aviation regulations. Moreover, the Aviation Training Centre, functioning as of 1976, operates as an independent extra-departmental unit of the Technical University. Another form of university activity is the AERONET – Aviation Valley (AERONET Centre for Advanced Technology) established in 2004. One of its activities involves the establishment of a Laboratory for Testing Materials and Materials Engineering equipped with modern research and scientific equipment for performing research on advanced "high tech" engineering solutions. Additionally, the Rzeszów Technical University Gliding Centre and Inter-University Multi-Purpose Research and Scientific Aviation Laboratory of the Technical Universities of Rzeszów and Warsaw have been operating in Bezmiechowa since 2004. The Rzeszów Technical University Aeroclub also (since 2005) runs intensive flying training courses.

Klaster LifeScience Kraków (LifeScience Cluster) – PL, Lesser Poland Voivodeship

Klaster LifeScience with its focal point on biotechnology and life sciences was established by 32 entities in 2006 in the Lesser Poland Voivodeship. With its main vision of fostering innovation processes and technology transfers in its sector, it associates specialised institutions focusing their scope of activity on different areas of life such as: biotechnology, pharmacy, medicine, food and environment; universities, institutions of higher learning and other educational units; research and scientific institutes; hospitals and foundations connected with health care; consultancy companies and other business support units; as well as local authorities. Currently, the cluster comprises 71 companies and other institutions.

The cluster is extremely active in the area of **networking** using different forms such as the organisation of networking events, for example, regular meetings and mini-conferences, workshops and training, for example the organisation of a "Life Science Open Space", where ideas for new projects are born and scientific and business contacts are developed (connect & develop notion). Additionally, Cluster Life Science has created its own blog in order to ensure the exchange of information among its members and keep them up to date with news regarding the cluster activity. It also uses an innovative form of marketing and PR, as it has created a discussion group on the well-known platform of LinkedIn, thus enhancing communication and collaboration with international experts in the area. Additionally, due to the division of entities according to thematic groups, the cluster ensures the best possible work environment, the facilitation of communication and the engagement of specialists in the particular area of the implementation of concrete aims. In addition, the cluster supports entrepreneurship and innovation in its field and creates conditions for the successful commercialisation of the results of research and development projects. It appointed a company in the USA in order to make possible new contacts with similar profile companies (integration of the life science environment), and facilitate the exchange of knowledge, experience and technology transfers. LifeScience Cluster supports **spin-off creation** (with new ideas and inventions) and **incubator services** as well. It also aims to implement various projects i.e. the database and knowledge base, the exchange of ideas and initiatives, innovation fairs; co-operation with Seed Capital institutions and Business Angels.

CREA Hydro&Energy (Czech Renewable Energy Alliance) - CZ, South Moravian Region

The beginnings of co-operation between companies creating the core of the cluster dates back to before 2000. In 2004, the Czech Renewable Energy Alliance was established and with the support of Czech Trade, marketing projects in the Philippines and Vietnam were jointly implemented by five associated companies (of which three currently create the cluster's nucleus). Upon completion of the projects, the association continued in their common activities and expanded their co-operation with further companies and educational institutions. In 2008, thanks to the initiative of the companies, the CREA Hydro & Energy cluster was established in Brno in the South Moravian Region with its main objective involving supporting innovation and increasing the competitiveness of its members. Currently, the cluster contains 15 members with a focus on special services for water and waste management and sustainable energy sources, and with additional international partners in the industry, for example, from Iraq, Serbia or Vietnam.

The cluster places a lot of emphasis on its **enhanced international co-operation**. It co-operates with large Czech as well as international corporations. It uses the latest technologies and know-how in its area, alongside its development and transfer within its international co-operation. CREA Hydro & Energy is additionally a partner of the Ministry of Foreign Affairs of the Czech Republic in the implementation of training for Iraq experts within the Aid for Trade programme as well as further projects focused on the development of foreign co-operation. Additionally, the cluster is a member of the Business Platform for International Co-operation Development, whose members are also members of international associations such as ICOLD (International Commission on Large Dams), ESHA (European Small Hydropower Association) and others. Within the CORNET project, it also co-operates with other clusters, for example, from Germany, Belgium and Slovenia. The cluster has also been implementing numerous international projects. One of them is in the Philippines, where the consortium of CREA Hydro & Energy and its members improved the original design of the mini hydropower Plant Sevilla on Bohol Island and were awarded a contract for the detail design, manufacturing, supply and installation of the electro-mechanical equipment in this plant. Further examples are international projects implemented in Iraq, where the cluster carries out the reconstruction, modernisation and extension of TBD (technical and safety supervision) monitoring systems on two newly constructed dams i.e. the Dokan Dam and Derbandikhan Dam, as well as works connected with an increase in the safety of water structures. In addition, the cluster implemented a project called the Panel of Consultants for the Ministry of Agriculture and Water Resources (MoAWR) in the Kurdistan Region Government, which is aimed at providing consulting services in the area of engineering geology, hydraulic engineering, hydraulics and geotechnics. In total, 159 reports concerning 18 locations in the south of Iraq were developed. During the Panel, complete projects of water structures were assessed, two dams were completed, construction of four dams were launched and projects for a further 12 dams were prepared. Non-member companies alongside cluster members took part in common projects in Kurdistan, in the form of subcontracting from the area of the water industry amongst others. Of great significance for the cluster is the CREA Company, which is the local representative in Iraq. Another project constitutes the Feasibility study & design for Bawanur Dam & Garmian irrigation canal in the Kurdistan Region. The first phase of this project covered the preparation of the feasibility study, while the second phase included detailed topographical works, hydrological studies and geological-surveillance fieldwork. The last phase covered the preparation of a project for an irrigation system for the Garmian region, including a proposal of variants of the irrigation system and the determination of areas suitable for irrigation.

Klastr Přesného Strojírenství Vysočina (KPSV, Precision Engineering Cluster Vysocina) – CZ, Vysočina Region

Klastr přesného strojírenství Vysočina was established in 2007 in Moravské Budějovice and currently is the only functioning cluster in the region. At present, KPSV comprises 20 member companies from the Vysočina and the South Moravia regions, and three educational institutions (Secondary School of Crafts and Services in Moravské Budějovice, Brno University of Technology and College of Polytechnics Jihlava), which primarily deal with machinery and locksmith production alongside precision machining. Its main objective is support, development and strengthening of competitiveness, economic growth and the long-term prosperity of companies in the industry, their exchange of information, R&D development and ensuring costs benefits.

The primary activity of the cluster is the implementation of joint **R&D projects** in the area of engineering; within 14 of the cluster's joint projects, 12 are in the area of R&D. The projects address areas such as: applied research, development and innovation based on the needs of cluster members, cluster promotion and human resource development, acquisition of testing, laboratory and measuring machines and equipment for the cluster, networking, know-how and capacity sharing, acquisition of hardware, software, networks and IT for the cluster.

Thanks to the joint co-operation of cluster members, a common product, a unique patent for the production of a roof tile made from a special mixture of crushed PET plastic, fine quartz sand and further components has been developed. Thanks to the indestructible, consistent integrity and high-quality, and durable materials of the plastic-concrete, roof tiles are being provided. This tile is characterised by unique parameters such as being four times stronger and up to 40% lighter than conventional roofing tiles, having a self-washing capability with rain and snow, zero water absorption preventing the formation of moss, and numerous others. It provides savings of wood materials in new buildings, lightweight trusses and weaker battens, as well as being particularly suitable for the reconstruction of older roofs as it eliminates the need for reinforcement of roof trusses and wide ties. The patent is for both the product and prototype device. It has already been patented in the Czech Republic, with an aim of also patenting it in the European Union, Asia and America. There has been an interest in the product from China, Hong Kong, or companies from Bulgaria and Ukraine.

Moravskoslezský dřevařský klastr (MSDK, Moravian Timber Cluster) – CZ, Moravian-Silesian Region

The Moravian Timber Cluster was founded in 2005 as a result of the project/research study, conducted by the Union for the Development of the Moravian-Silesian Region. Currently, the cluster comprises companies in the wood and construction industries, as well as representatives of higher education, totalling 30 entities. Its structure covers the entire supply value chain from wood growing, its subsequent processing to manufacturing products with high value added. MSDK's main goal is to innovate and develop activities that would improve conditions for doing business in the wood industry and strengthen links between research, universities and businesses. The cluster's key areas of interest include R&D, education, as well as marketing and promotion.

In the area of **joint R&D**, the cluster has already implemented a project entitled "Creation of a professional training centre for the Moravian Timber Cluster", co-financed from the Operational Programme Enterprise and Innovation, Training Centres programme. Its aim is to build a professional training centre within the premises of the Faculty of Civil

Engineering at the Technical University of Ostrava, intended for education and human resource development of the MSDK members. This centre serves as a unique and visual learning tool for passive low-energy wooden construction, where technologies in passive standard and environmentally friendly materials are presented through professional workshops, courses and conferences. The latest technologies were used for this passive house construction. The cluster continues the development in this direction, aiming at building a five-story wooden building. The application for the implementation of this project was submitted on 28th of June 2012, with a total budget of approximately CZK 70 million.

In the area of **human resources and education**, the cluster has already implemented projects such as “Improving quality in education” with the main objective being to increase the popularity of wood and construction specialisations with a focus on elementary and secondary schools in the region and improving the quality of their teaching with regard to the requirements of the MSDK members, or the “Further education of employees of schools and school facilities” with the main focus on the development and improvement of professional education of teachers in six technical high schools and vocational schools in the area of wood processing and construction. Additionally, the cluster is preparing for the implementation of subsequent similar projects.

In the area of **marketing and promotion**, the cluster is a successful organiser of events such as the Wood Festival, student project competitions, or the WoodRoadshow. The Wood Festival focuses on the general popularisation of the timber industry. It has both an educational and promotional character starting with demonstrations of forest machinery in action and exhibitions for older guests, up to teaching about the forest, nature, and animals, etc. for the youngest visitors. The student project competition, focused on designing in the specialised software for timber construction and interiors, is directed at high school and university students from the Czech Republic and the Slovak Republic, which compete in several categories such as interior design, building design, equipment design (chairs, furniture, stairs, etc.). The WoodRoadshow, thanks to its entertaining and informative character, aims at attracting young people from elementary schools to study at technical high schools with a focus on the wood and construction industry.

Plastikářský klastr (Plastics Cluster) - CZ, Zlin Region

The plastics industry in the Zlín Region is a sector of key economic importance with a long tradition. Besides the region’s high concentration of such companies, it has a university with fields focusing on the plastics industry, and other related infrastructure. The cluster was established in 2006 by 18 founding members. The number of members in the cluster is rising constantly, today numbering 26 members, of which 80% represent small and medium enterprises. A fundamental mission of the cluster is to create a communications environment for companies dealing in plastics. Its activities focus on:

- Active implementation of the interests of its members within the framework of regional policy in the areas of support of enterprise, innovation and education;
- Promotion of the plastics field;
- Building technical infrastructures for research, development and training of operators of plastic machines;
- Providing selected services for cluster members.

The first common project by the cluster was the **joint purchase** of electricity, which was successfully implemented and encouraged the implementation of further projects by cluster members.

The joint energy purchase began with purchasing electricity in 2007. Eight companies took part with a total electricity consumption of 18.4 GW·h. This resulted in decreasing the final price by CZK 1.74 million below the initial offer. This successful project continues with an annual declaration of joint demand management. For 2013, 19 companies have taken part in the joint electricity purchase with a total estimated consumption of 105 GW·h. Over the course of seven years, such negotiations have enabled a decrease in the margins of traders on the stock prices by 70%, to which accession of a company contributed in 2010 - one of the most important companies in the industry - which covered the joint purchase while simultaneously facilitating electricity purchases under even more advantageous conditions. For 2013, agreement was reached for a margin that is 57% lower than that of 2011. This margin decrease projects into the resulting price by decreasing the total price of the electricity commodity component by 4.4%. Today, the savings made by this common demand management can be quantified to approximately CZK 11 million annually.

A key to this successful common selection procedure for electricity purchase was common demand management for purchasing natural gas in 2010. Eleven companies took part, buying a volume of 22.2 GW·h and attaining a cost savings of 30%. Today, 13 companies are taking part in the common selection procedure for natural gas purchase with a volume of

23.5 GW·h. The negotiated margin decreased by 55.7%, where use of stock market spot prices also contributed. This margin decreased the projects into the resulting commodity price by decreasing the total price by 5.6%. Today, the savings made by this common demand management can be quantified to approximately CZK 1.8 million annually.

Automobilový klaster – západné Slovensko (Automotive cluster – West Slovakia) – SK, Trnava Region

As the first cluster in the Slovak Republic, the Automotive cluster - West Slovakia was founded in 2007 by the Trnava Municipality Region and the City of Trnava, which agreed on a shared approach to establish, build and manage a project with the same title, with its goal being to increase the competitiveness and quality of the production of the supply companies operating in the automotive industry as well creating qualified job positions. Currently, the cluster comprises 38 entities from the industry, education and R&D spheres.

The cluster is remarkably active in the area of **networking** and **human resources**. It characterises exceptional skills in partnership building with its enhanced co-operation activities conducted on an international scale within numerous projects such as Autoplast, Autoclusters, AutoNet, cross-border projects with Hungary (HU - SK "Klastre bez hraníc" – Clusters without borders) and Austria (AC Centrope), and others. These are networking events such as conferences, workshops, forums, matchmaking events, etc. with the aim of promoting the region and its actors in the automotive industry, boosting the creation of international co-operation, ensuring professionalism and expertise in the industry field, alongside the creation of awareness about supporting new innovative solutions and the automotive industry as a whole at various policy decision maker levels. In addition, the cluster also actively participates in trade fairs (for example Autosalon Incheba, CARplast, Wienna-tech) and events in the industry organised by others, thus through the **joint logo and advertisement** of the cluster, ensuring the recognition of its members and obtaining new contacts for business and project co-operation. Currently, the cluster co-operates with numerous clusters both within the Slovak Republic as well as abroad i.e. from Austria, Croatia, the Czech Republic, Hungary, Serbia and Slovenia, alongside other entities from Italy, Poland and Romania.

Additionally, the Automotive cluster - West Slovakia is extremely active in **lobbying** activities, particularly for favourable legal regulations. In 2012, it became a member of the Automotive Industry Association (AIA), where the main goal of co-operation is lobbying the government for improved utilisation of national resources and also direct access to and influence on legislative rules and norms which are currently in preparation or in progress in the Slovak Republic.

Styrian Cleantech Cluster ECO WORLD STYRIA (ECO Cluster) – AT, Styria Region

The beginnings of the development of the Cluster ECO World Styria date back to 1998. Since that time, the cluster, established in the World's Green Tech Valley in Austria, has been developing dynamically and today comprises 200 companies and research centres in the fields of renewable energy and environmental technology. The cluster's vision is to strengthen and develop the position of Styrian companies as international technology and market leaders and support the Styria region by driving innovation, know-how and the development of new markets.

The cluster is exceptionally active in the area of **networking** focusing its activities on initiation of co-operation, in particular in research collaborations, as well as supporting the internationalisation of cluster members. It provides well-organised information support (for example, through an Eco World Magazine presenting the latest news from the industry, thus keeping both domestic and foreign readers up-to-date), and which organises networking meetings, workshops, conferences, congresses and other events in order to ensure the acquisition of know-how, continuous development, consolidation and improvement in the sector as well as driving technology innovations. In addition, it offers a so-called ECO Future Radar, constituting a tool for strategic orientation developed in order to help established companies and newcomers familiarise themselves with details in the sector, market developments, top market trends, etc. In addition, the world-leading products and services of the cluster members are presented in a compact and accessible format for global customers in the Green Tech Valley Guide 2012, which is freely accessible. This, in turn, constitutes an excellent tool for broadening **business co-operation and promotion**, and meeting with growing international demand. Thanks to the guide, companies, local authorities, and private individuals can quickly and easily find appropriate products and services from the Green Tech Valley. Moreover, the ECO Cluster already co-operates with 35 other clusters from Austria and various European countries. It is additionally a partner in the "Eco-Innovation Cluster Partnership for Internationalisation and Growth (EcoCluP)", the first pan-European cluster partnership bringing together key European clusters focused on a variety of eco-innovative industries. Furthermore, the cluster was the first in Europe to be awarded partnership status by the Chinese development group TEDA (Tianjin Economic Technological Development Area) aspiring to be a nationwide pioneer in green technologies. The cluster also signed a working agreement for the expansion of business contacts with the China Solar Valley.

A strong focus on the development of technology **innovations** and boosting top-level **research** is an inseparable part of the ECO Cluster activity. Consequently, the proximity of research facilities such as, for example, the University of Leoben or Graz University of Technology constitutes an important advantage. The stimulation of thinking outside the box and inter-disciplinarity are realised with the use of concrete levers such as the Technology Round Table, CIC Cleantech Innovators Club, subject-related research (patent analysis, market research, etc.), funding maps, etc. The cluster has a great influence on research strategy for the region; it also supports its members with their search for project partners, exploring possible target markets, as well as helping with the search for suitable forms of finance.

Holzcluster Tirol (Tyrol's Wood Cluster) – AT, Tyrol Region

Tyrol's wood industry, rich in tradition, has become an innovative and high-tech economic sector over the past few decades.

The wood cluster was established in 2003 with support of the Tyrol provincial government (Land Tirol) and proHolz Tirol, a marketing organisation focused on supporting forestry and the timber industry. The cluster now has over 90 members, where these cluster members include the entire value chain of wood processing (from forestry to sawmills, the building industry, furniture production, etc.).

The main principle behind the cluster's work is supporting co-operation of its member enterprises in specific projects. These projects are then financed independently, and are operationally managed by the cluster. Amongst the successfully implemented activities of this cluster are its **marketing projects** "Zirbe" and "Holzfenster".

The "Zirbe" project focuses on using a special type of lumber from a tree that grows in Tyrol, known as "the Queen of the Alps" (Zirbe, Eng.: Stone or Cembran Pine; Lat.: pinus cembra). Research preceded the cluster's establishment aimed at analysing the properties of this wood from which emerged positive health effects of Stone Pine furniture. This type of wood, which is, et alias, resistant to attack by bacteria and moths, is highly popular in Austria, Germany and other nations where the wood has traditionally been used for centuries. The cluster's focus since 2004 has been to engage in marketing activities to increase awareness and consequent support in using this wood type. Brochures have been created describing the wood's properties, and all cluster members have them available to share. Other advertising materials and a separate website were also created.

The "Holzfenster" project includes the creation of a working group with the aim of implementing marketing and lobbying of the benefit of using wooden windows. Its main efforts concentrate on the promotion of government subsidies (preferential treatment) when using wooden windows as opposed to plastic ones. This is because wooden windows are more expensive than plastic ones, but their advantage is a far longer service life. They also have effects that positively influence employment (it takes many more employees to produce wooden windows than plastic ones). The project aim is also the exchange of experience and good practice amongst window producers. The cluster management organises and co-ordinates this work group's activity.

The presentation of best practice examples and sharing experience of active and successful clusters (with a history of existence) may serve as a learning tool for other initiatives which are at the very beginning of their development.

CHAPTER FOUR

KEY FACTORS FOR THE SUCCESSFUL CLUSTER DEVELOPMENT

Clusters, in order to become competitive in the global market or maintain their already existing competitive advantage, create a strategy and action plan, try to identify their opportunities, partners for co-operation and ensure long-term benefits such as implementing joint projects, etc. Their successful development may vary as numerous factors have influence on clustering processes. Thus, identification of the key critical factors for cluster development constitutes a guideline as to what to pay special attention to in order to become successful and a useful tool for the evaluation of the level of cluster excellence as an individual and in comparison to others (by using the method of benchmarking). The recognition of these factors appears as a reference point for the evaluation of the cluster development status and constitutes the basis for the formulation of the framework conditions for an ideal environment for the cluster growth.

Factors influencing cluster development

The development of clusters may be influenced both by external as well as internal factors. The external factors affecting the development of individual entities include the framework conditions at the macroeconomic level, public policies, national and regional strategies, legal regulations, etc. There are, in addition to these, also internal factors, with the micro-cluster perspective shaping the conditions for the functioning of individual clusters. They are influenced by the nature of these clusters, its structure, governance, set targets, etc.

Despite the fact that each cluster is unique, there are several features which can be identified and distinguished as common factors characterising a successful cluster. Among these are "soft" characteristics such as a network and "hard" characteristics such as a physical infrastructure, the presence of large companies in the cluster, or intangible aspects such as the existence of an entrepreneurial culture in the cluster, etc. There are also other factors such as access to markets, financing or specialised services.

Based on the critical literature review, the following factors were distinguished as ones which are referred to as extremely important, crucial, critical, etc. for the successful development of clusters:

- Networking;
- Mutual confidence of cluster members (trust);
- Mutual communication amongst companies;
- Personality of the cluster manager (director, facilitator);
- Professionalism of the cluster manager;
- Access for companies to information;
- Use of ICT/virtual media for communication;
- Presence of a large/dominant company in the cluster;
- Presence of a multinational company in the cluster;
- Education of human resources;
- Co-operation with educational institutions;
- Joint research/co-operation with a research institution;
- Innovative technology;
- The cluster covers entire or most of the value chain;
- Strong entrepreneurial spirit in companies;
- Competition rivalry amongst companies;
- Access to finances;
- Subsidies from the government during mapping and formation of the cluster;
- Subsidies from the government during growth of the cluster;
- Access to support services;
- Developed infrastructure;
- Monitoring of productivity of companies in the cluster;
- Benchmarking;
- Lobbying in favour of infrastructure, legislation, subsidies, etc.

The cluster is a complex network that is based on teamwork amongst its actors, not on their individual efforts. Therefore, one of the very first steps crucial for its development is **networking** and obtaining motivated and active participants for the cluster. These should be participants who understand the cluster concept and the value of the benefits that can be achieved thanks to this type of co-operation. Together, they should design a strategy and action plan for the cluster.

In order to find a common consensus in defining common goals and identification and further implementation of joint activities, a certain level of mutual **trust** among the participants has to be present. **The mutual confidence of the partnering entities** within the cluster is fundamental for its further development. However, as trust building is an on-going process, the formulation of the cluster on the basis of previous co-operation of certain members creates an ideal base for further networking.

Strengthening the formal and informal social networks is an indispensable element of a well-functioning cluster. Informal relationships and personal meetings allowing for the creation of long lasting communication channels among the particular actors often play a key role. **Mutual communication among members** is often facilitated by the cluster **facilitator** who is responsible for familiarising individual cluster participants, creating links, removing communication barriers (and any past conflicts between them), and convincing them of the cluster idea. Here it is vital to remember that clustering is a voluntary activity and that you cannot force someone to participate. Attempts at persuasion may result in the passiveness of those actors who enter into the cluster for the wrong reasons. The facilitator should address potential cluster members individually and personally as private discussions favour greater openness rather than shared meetings at, for example, public workshops. Adequate preparation on the part of the facilitator of the meeting (**professionalism**) is also equally important.

Another crucial task is to define the borders and participants of the cluster. Too broad of an industry focus for the cluster and too diversified a range of its participants may cause difficulties in finding common objectives and activities. Also, taking into consideration the geographical range of the cluster and administrative boundaries of the regions in which the cluster operates may be significant in regard to the support and involvement of regional authorities. Furthermore, as a lack of sufficient belief in success is typical for the early stage of the cluster development, the frequent passiveness of a large majority of cluster participants has been observed. It is therefore vital for the **leader of the cluster** to obtain members who will be active and enthusiastic about the entire clustering idea and who will push the cluster forward.

The leader may be the **cluster manager** or a key **large (dominant) company in the region, a multinational company**, or simply a company with rich resources and high innovative potential; respected, experienced, with recognition on the World markets, etc. The large company in the cluster constitutes a source of information and expertise (provides a critical mass of experienced managers and qualified workforce), technology (creates conditions for high innovation and technology growth and development), markets (creates the customer and supplier base), etc. Therefore, its key role can generally be seen in playing a catalytic role in various respects, for example, in disseminating and sharing knowledge and technologies among SMEs, encouraging innovation and reducing transaction costs, encouraging and inspiring future businesses, and spin-offs. Frequently it is the case that the cluster leader is the representative of a university (where the idea of the cluster is initiated). Regardless, the leader is always responsible for driving the development processes and stimulating the activity of cluster members.

The appropriate juxtaposition of the cluster steering group (cluster management) i.e. the choice of the cluster manager (facilitator), the creation of a project team led by the cluster manager, the selection of the board of directors and the supervisory board of the cluster, etc. is a key factor for the cluster and its successful development. As each cluster is specific, there are no precise instructions as to what the composition of the steering group should look like. It is important, however, to include representatives from all groups of stakeholders i.e. representatives from the industry (business) sphere, academia, research institutions, regional authorities, etc. The steering group should be composed in such a way so as to include all the competencies required for successful cluster development in the various phases of its life cycle. The **cluster manager or director** should have certain **personality characteristics** such as high qualifications, professionalism in his or her field, be a charismatic person who can spread energy and enthusiasm, motivate cluster members and ensure their active participation in joint initiatives.

The cluster leader is often a person associated with the region in which the cluster operates, with an excellent ability to develop relationships between cluster participants. In order to support networking within the cluster, the manager should ensure the presence of various tools such as: regular meetings of representatives of cluster members, mediation of contacts between cluster members and external organisations, **access to and information support** in different forms, for example, through the cluster website, the database of cluster actors, a cluster newsletter, and undertaking initiatives conducive to the improvement of communication channels (both formal and informal) among members (for example **use of virtual media**).

The cluster owes its success to its participants who, with the use of these different tools, create high-quality interpersonal relationships and strategic alliances characterised by a certain level of dialogue and trust. Here, social capital in the form of human resources appears as a vital factor for the development and performance of the cluster. **The education of human resources** plays a significant role as the presence of a skilled and dedicated workforce within the cluster structures acts as a magnet attracting further potential members (companies). The cluster may thus offer increased specialisation, availability of inputs and subcontractors. Additionally, the concentration of producers in a particular area attracts companies which provide specialised support services, while a large number of specialised suppliers and service providers contribute to numerous benefits such as a competitive price, prompt delivery and others. The cluster is also considered extremely attractive by customers when they can find products, and where the critical mass and the image the cluster creates opens access to markets on a local, national and international scale (access to a wide population and dynamic markets, internal and external markets, etc.). Strong social ties amongst the members enables a rapid flow of information and new technologies, makes them conducive to innovative (technology) development, while at the same time ensuring that they all remain in the cluster.

Universities and research institutions, being the source of human resources serving as generators of knowledge and new technologies (for example joint research), are essential partners in the cluster. Their quality and quantity is critical in developing the successful cluster. Moreover, the quality and availability of further education of human resources (the potential workforce) is also vital for cluster development. Clusters should therefore develop activities in order to deepen those already existing and establish further relations between industry and these organisations, thus ensuring their co-operation. Clusters, through their development may influence the quality and availability of further education so that the educational infrastructure will be able to respond to the requirements of the employers (flexibly matching supply and demand) and provide adequate professional preparation and training. The cluster itself may provide different activities in the area of human resources such as vocational counselling, consulting and information services (for individuals or companies) or the creation of special training centres, etc.

Another aspect of co-operation with these institutions is the **innovative (technology) development** of the cluster. Innovativeness determines whether the cluster is able to maintain its position on the market, while a strong research base is fundamental for the new ideas for its future development. Joint R&D activities lead to the development of new knowledge, which is the basis for the introduction of new technologies and innovation of products and processes. The cluster may support these activities through the networking of member companies (cluster actors) as well as building science and technology parks in order to attract business oriented research and thus create a highly technical business environment. Additionally, strengthening ties between research institutions and companies with specialised support service organisations (for example incubator services), the use of university laboratories and industrial needs for commercialisation of research (the possibility of spin-off creation) or creation of technology transfer institutions (centres) accelerate the transfer of the technology and innovation crucial for successful cluster development.

Building or strengthening the **links within the supply chain** between the companies in the cluster may constitute another important factor. Additionally, attracting investments and thus increasing the number of co-operating companies or filling the missing gap in the supply chain (where **the cluster covers the entire or most of the value chain**) may contribute to the successful growth of the cluster.

A growing number of companies (for example newly emerged, spin-offs) in the cluster which bring along fresh ideas and which are characterised by a **strong entrepreneurial spirit** are acknowledged as an important factor as well. Of significance for the cluster is that the number of individuals co-operating within its structure has a high ability to adapt to market changes, and is flexible and open to implementation of new things. They thereby benefit from new opportunities and technologies, bringing innovation to the market or countering potential risks in an easier way.

Additionally, clusters develop in the area where the **competition (rivalry among companies)** is present as it inspires, motivates and stimulates innovative activities and processes within the cluster.

Another important factor influencing cluster development significantly is **access to finances**. No cluster can successfully develop if it does not have sufficient financing. Therefore, the cluster facilitator or manager has to primarily focus his or her attention on the cluster's financial needs and how to fulfil them. Clusters and their activities can be **subsidised from public resources** on a national and/or a regional level in many countries, or from projects financed internationally, for example, within EU funds. However, the use of these sources should not be abused by the clusters, nor should their existence be conditioned by them. All types of subsidies should be treated as a form of acceleration or as an increase in the size of the projects implemented by the cluster. The cluster, in turn, should ensure appropriate financing of its activities from private resources. The high percentage of SMEs that form the cluster often means, however, that the resources of its members may not be enough to achieve their ambitious goals. It is therefore necessary to look for other possibilities in the form of, for example, credit from a bank, venture capital, acquiring a financially strong member or investor, attracting foreign investments, or building or extending contacts between companies and investors. The proximity and co-operation with financial institutions and intermediaries can also be beneficial for the cluster.

As the first stages of cluster formation and development are the most crucial, public **support in the form of subsidies from the government during mapping and formation** as well as **during the growth of the cluster** is extremely important. Public support (also in the form of support from EU Structural Funds) through providing a "cash injection" may constitute an incentive in the cluster initiative development, and may be used to promote mutual co-operation for identifying clusters, etc. Although public support may play an important role in the early stages of cluster development, it is essential after a certain time that the industry takes over the financial burden and ensures the financing of the cluster activity. The ability to raise adequate financial resources which enable the implementation or expansion of the planned activities and ensures the long-term development of the cluster, its stability and financial independence, determines its success. These clusters are then able to achieve more ambitious objectives.

Access to support services may also be significant for cluster development such as support for new companies, spin-offs, consultancy in various matters (for example banking services; insurance, legal, accounting and tax consultancy), ICT support, assistance in obtaining grants, business advice on management, production methods, training, planning, marketing, etc. Support services may bring benefits in terms of the development of the corporate base, growth of new enterprises, and thus influence economic development.

Additionally, a **developed infrastructure**, i.e. available communication channels, a technical infrastructure, etc. constitutes a key factor for the development of the cluster. A high quality technical infrastructure may lead to reduction of costs (for example, in transport and in production), enable improved access to raw materials, other materials and a skilled workforce, improve communication between suppliers and customers, and thus bring financial benefits for the cluster and its members.

Monitoring of productivity of companies co-operating within its structure and comparison of the cluster with others operating in the same industry in the form of **benchmarking** may also be vital for cluster development. The cluster is consequently able to identify problematic areas or the under-performance of particular members and thus introduce necessary steps to improve the situation and secure its development.

Last but not least, the existence of a supportive policy, stable macroeconomic environment creating a background for optimised development of the cluster and its members is important. **Lobbying in favour of legislation, subsidies, infrastructure**, etc. is therefore extremely significant. The following can be distinguished among the conducive political factors: an economic policy aimed at removing barriers for competition; support for science and new technologies by the government; stable and predictable decision-making of the government; decentralised decision-making on a regional level; investments in the improvement of the transport and technical infrastructure; the existence of investment incentives; and trust between business and government bodies, etc.

Apart from key factors influencing successful cluster development, there are also those which may hamper influence and lead to the cluster's failure, such as the isolation of cluster initiatives, lack of a broader consensus on key factors of economic performance, lack of a strong network among cluster actors, an inadequate budget for the cluster activity, its sustainability issues, neglected brand building, and others.

Opinions of cluster managers on key factors conducive to successful cluster development in the selected countries (Poland, Czech Republic, Slovak Republic, Austria)

The conducted research was, amongst others, oriented on identifying **factors essential for successful development of clusters**. Based on a study of literature and various studies dealing with the issue of clusters, factors were identified that appeared most frequently in these resources and were deemed very important, crucial, critical, etc., for the successful development of clusters.

Within the questionnaire survey and structured interviews, representatives of the managing groups of clusters were asked to assign a weight to these factors. The results are illustrated in Table 4-1.

The analysis of the results obtained allowed classifying them into the following four groups (Table 4-1):

a) Critical factors (marked in black);
b) Significant factors (marked in dark grey);
c) Factors with average significance (marked light grey);
d) Insignificant factors (marked in white).

Table 4-1 Factors critical for a successful cluster development according to the opinion of clusters managers from the selected countries

Critical factor	Average value	PL	CZ	SK	AT
Mutual confidence of cluster members	4,73	4,76	4,78	4,67	4,50
Mutual communication amongst companies	4,65	4,67	4,68	5,00	4,29
Personality of the cluster's manager (director)	4,57	4,54	4,65	4,63	4,50
Access to finances	4,51	4,69	4,28	4,56	4,50
Networking	4,49	4,44	4,41	4,56	4,86
Professionalism of the cluster's manager	4,35	4,35	4,28	4,00	4,79
Strong entrepreneurial spirit in companies	4,33	4,37	4,31	4,44	4,15
Subsidies from the government during growth of the cluster	4,29	4,45	4,03	4,56	4,31
Joint research/co-operation with a research institution	4,21	4,24	4,35	3,78	4,00
Education of human resources	4,21	4,37	4,28	4,22	3,38
Access for companies to information	4,15	4,37	3,98	3,89	4,00
Innovative technology	4,12	4,25	3,90	4,33	4,15
Subsidies from the government during mapping and formation of the cluster	4,11	4,08	3,90	4,25	4,77
Co-operation with educational institutions	4,04	4,14	4,08	4,33	3,31
Lobbying in favour of infrastructure, legislation, subsidies, etc.	3,76	3,76	3,70	4,38	3,54
The cluster covers the entire or most of the value chain	3,42	3,59	3,13	3,13	3,85
Access to support services	3,41	3,56	3,13	2,78	4,15
Use of ICT/virtual media for communication	3,34	3,64	3,13	4,00	2,43
Developed infrastructure	3,28	3,65	2,67	3,78	3,31
Monitoring of productivity of companies in the cluster	3,14	3,51	2,65	3,56	2,92
Benchmarking	3,13	3,47	2,68	3,33	3,08
Presence of a large/dominant company in the cluster	2,82	2,76	2,85	3,67	2,43
Competition rivarly amongst companies	2,80	3,04	2,36	2,50	3,38
Presence of a multinational company in the cluster	1,99	2,02	1,93	2,56	1,71

Source: Own development
Note: Individual factors were evaluated by cluster managers on a scale from 1 to 5, where 1 means insignificant factor and 5 very significant factor. According to the weighted average the factors have been divided into four categories: 1) Critical factors; 2) Significant factors; 3) Factors with average significance; 4) Insignificant factors.

In general, it may be stated that managers of clusters from the analysed countries consider as critical for successful development the following factors:

- Mutual confidence of cluster members;
- Mutual communication amongst companies;
- Personality of the cluster's manager (director); and
- Access to finances.

Significant for development of the cluster are, in the opinions of cluster managers, the following factors:

- Networking;
- Professionalism of the cluster's manager;
- Strong entrepreneurial spirit in companies;
- Subsidies from the government during growth of the cluster;
- Joint research/co-operation with a research institution;
- Education of human resources;
- Access for companies to information;
- Innovative technology;
- Subsidies from the government during mapping and formation of the cluster;
- Co-operation with educational institutions;
- Lobbying in favour of infrastructure, legislation, subsidies, etc.

It is clear that for a cluster's development, mutual relationships amongst the cluster members are key. These are also supported by functioning networking, member access to information and a strong entrepreneurial "spirit" in the member companies.

Successful clusters often characterise strongly embedded networks, where mutual trust and interpersonal relationships are highly developed, and where social capital of the cluster is high as well. The character of the cluster network is under a process of continuous creation, and changes accordingly to the adjustments in the membership and needs of the cluster. Therefore, strengthening the links between cluster members is one of the most important elements of any clusters development strategies.

The cluster's management is a second key element for developing the cluster. Besides the very personalities of the cluster managers, another highly important aspect is their professionalism and experience in managing such a complicated entity.

Successful clusters are managed by the steering group that is characterised by a high level of competence and expertise. They are able to create a strong knowledge base in the sense of a higher level of knowledge management and skilled labour. This, in turn, is considered a key factor in attracting and keeping businesses as well as for the successful development of companies in the cluster.

The third aspect strongly influencing how the cluster functions is securing financing for its activities. Of course various options contribute to securing financing by means of subsidies, which clusters can obtain for building and further development from various sources, for example, from the European Union, from national or regional budgets.

The assurance of adequate financing for the cluster, especially during the early phase of its development, plays a crucial role for the formation of the entire network and for boosting its start-up activities. As it is a learning and trust building phase, it is always very difficult to finance, thus support in the form of external financing may be stimulating and in some cases even necessary. In the phases of further growth, the cluster is expected to become self-sustainable.

Case from the Czech Republic

Amongst the analysed clusters in the Czech Republic, 62.5% used one of the subsidies to create and develop their cluster, provided within the framework of EU structural programmes. In terms of the opinion on the importance of financial support of clusters from public sources (Table 4-2), most cluster managers in the Czech Republic (57.5%) feel that without financial support from the state, the existence of the cluster would be possible, but with a completely different scope and nature of activities. A large portion of managers (45%) is also of the opinion that state financial support is indeed necessary, but it attracts enterprises that only join the cluster with a flawed vision of obtaining funding. A third of those asked even hold the opinion that without state financial support, the existence of a cluster is not at all possible.

It is interesting to observe the opinions of managers in relation to whether or not their cluster capitalised on the opportunity to obtain a subsidy from European funds. Managers of those clusters that received additional funding are mostly convinced clusters' existence without financial support from the government is possible but it is then reflected in a different nature and scope of activities. They often believe in the necessity of that type of support even though it may attract enterprises to join the cluster for the wrong reasons. On the contrary, it is surprising that managers of clusters not having received these subsidies think that without state financial support, the cluster could not exist.

Table 4-2 The importance of financial support for clusters in the opinion of cluster managers

The importance of financial support for clusters in the opinion of cluster managers	All clusters without division	Clusters supported by subsidies	Clusters without subsidies
Without the financial support from the government the existence of clusters in not at all possible	30%	24%	40%
Without the financial support from the government the existence of clusters is possible, but to a different extent and activities	57.5%	76%	26.7%
Financial support from the government is necessary, but it also attracts companies which enter cluster only for gaining financial means	45%	52%	33.3%
Clusters should be a voluntary activity, financial support from the government is not necessary, companies would form clusters even without such support	7.5%	8%	6.7%
Financial support from the government is a negative element in the clustering process	12.5%	4%	26.7%

Source: Own development

The aforementioned factors are complemented by other key factors relating to activities on which the clusters focus. One such significant factor is co-operation with research and educational institutes and the consequent focus on developing human resources.

Lobbying also supports successful development of the cluster, whether for the benefit of gaining subsidies or legal changes, or the benefit of the infrastructure, etc. Managers of clusters also rated factors "with average significance", which are not considered to be critical for developing clusters. Of course it is always necessary to remember the fact that each cluster is unique, thus aspects that influence its successful functioning may be different and critical just in the specific conditions of a single cluster or a single economic environment. These factors include:

- That the cluster covers the entire or most of the value chain;
- Access to support services;
- Use of ICT/virtual media for communication;
- Developed infrastructure;
- Monitoring of productivity of companies in the cluster;
- Benchmarking;
- Presence of a large/dominant company in the cluster;
- Competitive rivalry among companies.

Only one from the evaluated factors i.e. presence of a multinational company in the cluster is considered as insignificant for a successful cluster development.

From the analysis performed it is clear that for the development of clusters in a specific country certain factors may be more important than others.

As it can be seen in Table 4-1, concerning **critical factors** of a successful development of clusters, opinions of the managers from all four analysed countries are quite similar.

When evaluating this group of critical factors, an interesting fact appears in that the Austrian clusters consider these factors as critical, with less significance than in other countries. In fact, they consider the factor "Mutual communication among companies" to be only significant, and not critical for the successful functioning of the cluster. On the contrary, in the case of Austria one may indicate other factors that cluster managers consider absolutely critical, and even with a greater value than the aforementioned four factors, these are:

- Networking;
- Professionalism of the cluster's manager;
- Subsidies from the government during mapping and formation of the cluster.

Slovak cluster managers have similar opinions, and also place huge emphasis on functioning networking and the opportunity to gain subsidies (specifically for growth and development of the cluster). Again, the importance and need for the development of the cluster policy and respective measures oriented on cluster support in the country are highlighted.

While focusing on the **factors important** for the successful development of clusters which were depicted in the above table, it is necessary to mention that their significance differs only slightly in individual countries. In the case of Austrian clusters it is surprising that cluster managers deemed education of human resources and co-operation with educational institutions as the factors with average influence. At the same time, Austrian clusters relatively strongly support human resources development and co-operation with educational institutions (for more see chapter three).

In addition, the Austrian cluster managers classify the "Access to support services" and the fact that "The cluster covers the entire or most of the value chain" among very significant factors.

The remaining analysed factors that could influence successful functioning of clusters, managers indicated as **factors with average influence**, and one factor as not significant at all.

Opinions of cluster managers on the most important achievements by clusters in the selected countries (Poland, Czech Republic, Slovak Republic, Austria)

The level of importance of particular factors for successful cluster development has also been highlighted within the conducted research when attention has been paid to the *most important achievements* which the cluster has already accomplished as well as the most serious problems encountered by the clusters.

The managers of **Polish clusters** emphasised the following among the most important achievements already accomplished by clusters:

- Building co-operation ties;
- Joint marketing and promotion;
- Information and education;
- Joint projects;
- Business co-operation;
- Financing.

The managers also distinguished **financial barriers** as being amongst the most serious problems encountered by clusters and the connected legal problems, co-operation issues and other issues.

Polish clusters contend with financial barriers such as:

- Lack of or limited financial resources (including external financing), own cluster resources (financial weakness, capital barriers) - inability to invest in projects, build their own infrastructure in the cluster, finance their own development, promotional activities, etc.;
- Problems with financing cluster workers – need for volunteering, need to be engaged in the cluster organisation on the basis of social work, lack of financial resources for the employment of a cluster manager;
- Incompatibility of EU funds (ignorance by the authorities of the cluster's needs), limited opportunities to use grants from the EU (cluster members can use certain sources while the cluster as a separate entity cannot) and the lengthy procedures for applying for EU funds (their difficult availability);
- Lack of sufficient financial support from the government;
- The bureaucracy and fortification of documents in applying for financial support;
- Impossible or difficult to meet requirements for applying for grants;
- Strong mismatching of funding instruments for the cluster needs, and for the needs of the cluster participants;
- Lack or limited financing on the regional level, for example, within the Regional Operational Programmes;
- Financial support "de minimis";
- Lack of funding at the beginning of cluster activity (financial support in the first stage of the cluster initiative);
- Lack of resources for supporting already existing initiatives.

Financial barriers are linked to **legal problems** where a lack of relevant legislation and regulations in regard to cluster initiatives in Poland has been observed. Additionally, difficulty in finding an appropriate model for benefit sharing among partners in the consortia (a number of ideas for projects, but at the same time existing problems in compliance with commercial law, for example, after the manufacturing of the product problems with the ownership) should be highlighted. Polish clusters also often face a lack of interest (passivity) from local government units, support from local and regional authorities, or even the government (ignorance on the part of all kinds of authorities regarding what clusters actually are and why they exist); and lack of willingness to co-operate on the side of training institutions, and institutions set up to help (only verbal support). They also face additional co-operation issues of an internal character (for example distrust, communication barriers) and issues such as insufficient municipal, governmental infrastructure or insufficient awareness on the part of the public in connection with cluster product/services etc.

The significance of particular factors for successful cluster development has been presented in Table 4-1, where mutual communication amongst companies and the mutual confidence of cluster members as well as access to finances and the personality of the cluster manager (director) have been underlined as the most important factors by the Polish cluster managers.

The managers of **Czech clusters** highlighted the following as among the most important achievements already accomplished:

- Joint projects and their successful implementation,
- Cluster establishment and obtaining of financial support,
- Joint presentation and advertisements,
- Structure and co-operation ties,
- Business co-operation.

The managers point out the importance of preparation and implementation of common projects as well as the cluster formation and their ability to apply for financial support for common project implementation.

This also constitutes the most serious barrier hampering cluster development in the Czech Republic. Clusters face administrative and legislative difficulties with regard to functioning as a cluster (the formality of cluster establishment in the Czech Republic) and the possibilities for applying for financial support connected with it. Clusters encounter the following **barriers** concerning **financing**:

- Lack or limited financial resources for cluster functioning and/or project implementation;
- No State support (subsidies) for establishing (starting) the cluster;
- Lack of resources for preparation of projects (pre-financing);
- Problematic funding from grants (difficulties in the fulfilment of the conditions in the support programmes; extensive administration of projects - bureaucracy; changes to rules during the execution of the project; legislative problems in regard to the project implementation - eligible costs, methodology of co-financing; a long period of

administration of the application of payment; late payments for the stages of the project from CzechInvest and the Ministry);
- Administrative problems, for example, when applying for loans (it is difficult to obtain loans from the bank without strong members in the cluster).

In addition, the distrust of members in regard to financing issues such as co-financing of investments within the cluster (companies do not understand that it is for them) is also inseparably connected.

Apart from financial issues, clusters face difficulties with **co-operation issues** of an internal character such as limited trust amongst members and distrust of potential cluster members, involving a reserve in co-operation and communication; apathy and unwillingness of companies to collaborate, as well as lack of time and therefore their low involvement in cluster activity.

Attention must be paid as well to the lack of or limited communication with regional authorities/institutions, and their lack of interest and deformed understanding of what the cluster is. In addition, poorly defined conditions and rules for cluster functioning by CzechInvest and the Ministry and a lack of concrete activities from the side of the Ministry for future development of clusters in the country has to be emphasised. External factors hampering cluster development include a complex tax environment, tax problems, the influence of the economic crisis and a lack of attention to a particular industry and/or service sector in a specified region (for example, the tourism concept in the Moravian-Silesian Region) have to be singled out.

Among the factors critical for successful cluster development in the Czech Republic, those connected with communication and collaboration such as mutual confidence and mutual communication amongst companies and the personality of the cluster manager should be highlighted.

The level of importance of the particular factors is also underlined in the fact that among the most significant achievements already accomplished by the analysed group of **Slovak clusters**, managers distinguish the following activities:

- Education and training;
- Joint trade names and promotion;
- Financing and networking;
- R&D;
- Joint products and projects.

In contrast, the most problematic issues on the part of Slovak clusters involve encountering certain difficulties regarding **networking** activities and **financing**. According to the managers, problems with financing of cluster activities constitute the biggest barrier hampering cluster development in the Slovak Republic. Clusters face difficulties with financing their start-up activities and finding resources for the professionalisation of the cluster. With no financial support from the government, clusters need to rely on self-financing (all the clusters collect regular membership fees from their members) which is difficult to obtain and which often results in keeping the costs of maintaining the cluster on a minimum level. In addition to this, a lack of legislation concerning the establishment and operation of clusters in the country, low interest on the part of the local government to participate in the activities of the cluster (no financial as well as information support) and the influence of the world economic crisis and the associated downtime of customer orders (and as a result a lower income), should be emphasised. Managers also point out problems connected with networking and co-operation among members such as distrust and the high workload of members, difficulties with finding a consensus among members, a lack of understanding of the clustering idea, and thus lack of involvement by future participants due to mutual personal conflicts with the existing members in the cluster.

The most significant factors critical for successful cluster development were, according to the managers of an analysed group of Slovak clusters, mutual communication and the confidence of cluster members, the personality of the cluster manager (director), as well as access to finances in general and support by means of subsidies from the government/region particularly during the growth of clusters; highest ranked in importance is networking.

The analysed group of **Austrian clusters** emphasises the following among their greatest achievements:

- Joint projects and their successful implementation;
- Creation of a platform for a comprehensive information exchange and communication;
- Training of entrepreneurs;
- Joint promotion (increase in visibility);
- Positive influence on the region and its development;
- Building a critical mass (engagement of a large amount of members).

The most serious problems encountered by the managers of the Austrian clusters involve **financing** issues, **the diverse corporate landscape** and **political interference**. Regarding financing clusters primarily face difficulties connected with reduced funding, the structure of financing and/or a decrease in interest in co-financing of cluster activity by entrepreneurs/cluster members and thus a lack or limited financial flexibility. This, in turn, is connected with the unwillingness of companies to co-operate and their apathy in action, their often parochial thinking and ignorance thus indicates a lack of understanding that joint efforts can result in becoming stronger and more competitive not only for an individual but also for the entire cluster. Another barrier consists of the fragmentation of the corporate landscape and the connected difficulties with the integration of the industries in which the clusters operate. The political influence in Austria involves other problems encountered. Clusters point out political interference as a negative factor particularly at the beginning of cluster development, where its growth is partially hampered by the public sector.

Amongst the Austrian clusters, networking and the mutual confidence of the cluster members, professionalism as well as the personality of the cluster manager alongside subsidies from the government during the mapping and formation of the cluster and with access to finances in general, are considered critical factors conditioning successful cluster development in the country.

Clusters as complex networks create a unique environment for the development of their member entities. When analysing key factors conducive to the successful development of the cluster, simultaneous attention has been given to its most significant achievements as well as to the most serious problems encountered. It is apparent that both the positive results of cluster activity as well as the difficulties they face confirm the level of importance and are consistent with the factors considered as critical for the successful development of clusters.

Benchmarking as a tool for performance measurement of clusters and their members

The growing pressure on competitiveness of an enterprise in an environment of market globalisation is compelling managers to assess and manage their business activities with a view towards constantly increasing company performance and strengthening its market position. Company performance and its measurement and control are becoming increasingly important in today's economic crisis and beyond. A handy tool for assessing the performance of companies in light of their proven capabilities to improve company results is **benchmarking,** a process of comparison with other organisations, on the basis of which the company implements activities leading to successful development. In practice, however, the performance of benchmarking encounters a lack of both information sources and partners for comparison.

This section introduces the results of co-operation of a cluster with a research organisation (this specifically concerns co-operation of the Zlín Region's Plastics Cluster and the Faculty of Management and Economics of Tomas Bata University in Zlín). The result of this co-operation was the creation of a benchmarking model for performance assessment that applies the potential for co-operation of companies in clusters. Based on the created model, management is able to uncover weak points in company performance when compared to others, and consequently adopt measures leading to improving their existing performance and market position.

The essence of benchmarking

Definition of the term "benchmarking".

In the area of improving quality, the term **"benchmark"** means achieving the level of "best-in-class". Achieving such a level becomes a reference point and a standard of excellence over similar performances or processes, which we measure and compare. While a benchmark is a comparative metric, the term **"benchmarking"** means a process of measuring and comparing against the given benchmark. We find a series of varying definitions of benchmarking in literature.

As Robert C. Camp (1989: 10) states, apparently the most popular definition comes from David T. Kearns, former CEO of Xerox, who stood at the birth of the systematic implementation of benchmarking. He defines benchmarking as the:

> ..continuous process of measuring products, services, and practices against the toughest competitors or those companies recognised as industry leaders.

No less popular is the definition of one of the leading figures in the area of benchmarking, Robert C. Camp (1989: XI), according to whom this process is founded upon "searching for industry best practices that will lead to superior performance of a company."

The concept of benchmarking has been defined by the International Benchmarking Clearinghouse of the American Productivity and Quality Centre (APQC, 1993: 4) as:

> …a process of continuously measuring and comparing an organisation's business processes against business leaders anywhere in the world to gain information which will help the organisation take action to improve its performance.

The Swedish consultants systematic Karlöf and Östblom (1995:9) also researched the process of comparison with other subjects. They define benchmarking as:

> ...a continuous, systematic process for comparing your own efficiency in terms of productivity, quality and practices with those companies and organisations that represent excellence. Benchmarking represents an excellent way of searching for and achieving success based on organic growth, i.e. its own performance-based growth.

The Ontario Municipal CAO's Benchmarking Initiative (OMBI, 2001: 12) quite generally defines benchmarking as:

> ....a procedure by which we are humble enough to admit that somebody else is better, and then we are smart enough to find out how we can match them or even surpass them.

From all these definitions we find that benchmarking relates to the necessity to look around (outside one's own organisation, branch, region or country) in the effort of searching for **best practices**, which Watson (2007:15) defines as "methods, approaches or activities leading to exceptional performance." It is then necessary to uncover the way of achieving subjects of these best practices and implementing them in one's organisation. If the findings ascertained through benchmarking are correctly applied, it is easier to achieve higher performance within the organisation.

Approaches to benchmarking

It is possible to compare practically everything, even in the everyday running of the organisation. According to what the subject of comparison is, the APQC (1993) distinguishes two basic approaches to benchmarking: performance and process.

Performance benchmarking compares relative performance amongst competitors or individual company units. In essence, it resolves the question: "What can the organisation achieve, what kind of performance?" No visits are required for such benchmarking; often it is performed as so-called "consortium benchmarking", where multiple organisations perform benchmarking - benchmarking agencies or centres.

Amongst the main advantages of this type of benchmarking is that it is a useful way of identifying the difference in performance amongst entities performing benchmarking, but also that performance benchmarking shows just where the organisation needs the most improvement.

The disadvantage of performance benchmarking is the fact that it usually does not identify processes or practices that lead to outstanding performance, and thus does not provide an "instruction manual" for the organisation on how to improve. It may also be difficult to interpret the results.

Process benchmarking measures the individual performance of the process and its functionality. Measuring and comparing generally takes place against organisations that are the leaders (the best) in performing specific comparative processes. Whereas performance benchmarking answers the question: "What performance does the organisation achieve?" the process concept of benchmarking searches for an answer to the question: "How does the organisation achieve excellent performance?" From this we find that process benchmarking ordinarily requires on-site visits of partners, thorough preparation and certain rules.

The advantage to process benchmarking is finding out how to improve and what level of improvement is feasible. It also makes it possible to interpret better and more easily the differences between compared organisations (why do organisations have differing performance).

The danger of benchmarking processes may appear if the organisation does not know what to compare or chooses a partner ill-suited for benchmarking.

Types of benchmarking

One of the important differences prior to starting benchmarking is choosing the right partner for comparison. From the aspect of potential partners, Karlöf and Östblom (1995) distinguish three basic types of benchmarking: internal, external and functional.

Internal benchmarking includes a detailed study of one's own organisation. This concerns analysis and comparison of selected activities, procedures or performances in individual departments, branches or affiliates performing similar activities with the aim of defining the best of them.

External benchmarking is a comparison with another organisation, with competition of its own market or equivalent organisation supplying other markets.

Functional benchmarking is a comparison of similar functions and methods with other top companies regardless of the field in which they function. If benchmarks of other branches are used, it is not possible to perform a comparison on an entirely inclusive level. Rather, this involves benchmarking individual parts of activities (for example work procedures) that demonstrate logical similarity in differing branches as well. An example may be the assessment of a supplier system in organisations in different lines of business.

In terms of the level of focus, Watson (1993) and Camp (1995) differentiate between strategic and operative benchmarking.

Strategic benchmarking focuses on questions regarding the company's long-term market presence, for example, building key competencies for maintaining a competitive advantage, development of strategically successful products and services, defining new markets, development of technologies, etc. Its result is the revision of the mission, vision and strategy for long-term sustainability and further development of its market position.

Operative benchmarking focuses on the understanding of specific customer requirements, gaining and applying best practices to their satisfaction by means of streamlining internal processes, thereby increasing the key company performance factors.

In terms of continual increase in the organisation's performance, it is appropriate to combine both approaches.

Benefits of benchmarking

Benchmarking is a process used by the private sector (and by the public sector in recent years), which supports managed change by means of an objective assessment of its own performance. This involves finding gaps between an organisation's own performance and the performance of the benchmark, which should evoke a feeling of dissatisfaction with the way things are functioning in the organisation, and the need for change.

Camp (1989 and 1995), APQC (1993), Karlöf and Östblom (1995), Andersen and Pettersen (1996), OMBI (2001) and Watson (1993 and 2007) summarise the **benefits of benchmarking**, which culminate in growth in performance and strengthening the competitiveness of companies, as follows:

- It enables companies to better identify their current market position,
- It provides timely warning if the company is lagging behind in its field,
- It provides solid information on weak and strong points within the company,
- It leads to determining more difficult objectives,
- Upon repeated comparison, it enables the constant achieving of better results,
- It helps companies focus on building a strategic advantage,
- It helps eliminate or exclude pointless activities, which leads to cost savings,
- It leads to better and faster decision making founded upon better information,
- It may highlight the changing needs of customers,
- It puts company strategies to the test,
- It introduces new ideas into the company and facilitates the sharing of experiences,
- It contributes to developing human potential towards higher independence, towards willingness to learn and change things, and becomes a motivational element.

Benchmarking is a systematic process aimed at searching for excellence within one's own organisation or in the surrounding world with the aim of gaining knowledge and adopting measures that make it possible to elevate the organisation to a first-class level.

Benchmarking slowly starts to be implemented towards clusters, where not only the companies within one cluster network are evaluated but also clusters among themselves.

However, one must remember that benchmarking, just like other management tools, also has inherent weaknesses. The user of benchmarking should also be informed on wide-ranging opportunities and threats, which may occur during the course of performing benchmarking.

Options for performing benchmarking within clusters

The existence of clusters brings with it new opportunities for assessing performance of enterprises, mainly in the form of a set of suitable benchmarking partners and easier access to data necessary for comparison.

As much research has shown, the main barriers to implementing benchmarking in practice are the lack of information sources and of benchmarking partners.

Clusters, as geographically concentrated groups of mutually co-operating companies (often SMEs), but which oftentimes also compete, represent the optimum opportunity for gaining data for the purpose of performing benchmarking. Clusters facilitate the exchange of knowledge and experience between the companies in the same sector through vertical or horizontal co-operation strengthening competitiveness of the undertakings. The average performance of the cluster as a whole or only a selected group of best companies may be a suitable benchmark for assessing the overall performance of a specific enterprise and for determining its position within the cluster with a view to increasing its market competitiveness. At the same time, the compared enterprise may be either a member of the cluster or an entity in the same field standing outside the cluster.

Despite the fact that benchmarking is spread across the business environment, it is not so much known or used by SMEs (Cassell et al. 2001). According to Broderick et al. (2010), most studies of benchmarking have identified and evaluated practices in large organisations. McCarthy (2003) notes the low number of benchmarking studies in SMEs; while McAdam and O´Neill (1999) identify the inapplicability of some benchmarking approaches for the idiosyncrasies of the small business environment. The problems of using benchmarking in SMEs mainly for cultural change, which is not easy to achieve, are mentioned in the work of McKay and Chung (2005) and Hwang and Lockwood (2006). According to Moriarty (2011), the relative simplicity of SMEs confers greater from practice-based frameworks than that commonly achieved with larger organisations. They can have common relationships, similar processes, and similar critical success factors.

As a common proactive measure, clusters may implement benchmarking between its members to overcome the insufficiency of information. According to Carpinetti and Oiko (2007 and 2008) or Škodáková et al. (2008), a benchmarking database can be a powerful instrument for information sharing and promotion of joint actions among firms within the cluster so as to improve organisational learning and the collective efficiency of the cluster. It might be worth considering launching a benchmarking database to fuel the exchange of information between individual firms and between clusters as well, perhaps on a governmental or even supranational level.

It is necessary to realise that properly performed benchmarking may provide a great benefit for not only the enterprises themselves, but also for the entire business group. Management of the group of enterprises normally wants to know if it is managing the cluster effectively, if it has correctly formulated its strategy and aims, and if they are being implemented in an appropriate way. In the event of state support, the question arises as to whether the support of business groups is beneficial and brings forth the desired effects. Not the least of which, it is also necessary to find out if the effects were achieved by reasonable means.

Applying benchmarking in relation to the origin and development of clusters can be summarised into the following areas:

- Assessing the performance of individual members of the cluster,
- Assessing the influence of bringing an enterprise into the cluster, both from the aspect of the enterprise itself and of the entire group,
- Assessment of the performance of the cluster as a whole,
- Assessment of the effectiveness of individual activities implemented in the cluster,
- Assessment of the effectiveness of cluster management (cluster initiatives),
- Assessing the influence of founding a cluster at the regional/state level,
- Assessing the effectiveness of support of clusters on the part of the region/state.

The relatively short period of existence of clusters, often their uniqueness and various specifics, have not yet enabled the development and creation of a definite and generally accepted manner of assessing any of the listed areas. On the other hand, one cannot say that no activities are being developed in the given area.

As stated by Pavelková et al. (2009), apparently the most important model used for assessing competitiveness of the company/region/economy is **Porter's diamond** (Porter 1990). Other important initiatives and projects dealing with the proposal of a method for assessing the stated areas in relation to the origin and development of clusters include:

- "Performance model of cluster initiatives", which proposes monitoring the performance of the cluster initiative in three dimensions (see "The Cluster Initiative Greenbook" 2003).
- "Initiative Cluster Linked over Europe", which lists indicators for benchmarking cluster initiatives (see CLOE 2006).

- "The Cluster Mapping Project"[1] and "The Cluster Meta-Study"[2], extensive research mapping clusters operating in the USA and in the rest of the world, performed by a group associated at the Harvard Business School entitled the Institute for Strategy and Competitiveness.
- "The Cluster Benchmarking Project" implemented by the Nordic Innovation Centre in Norway with the aim of developing a standardised tool for analysing cluster performance (see Andersen, Bjerre and Hanson 2006).
- The publication "A Practical Guide to Cluster Development"[3] from Britain's Department of Trade and Industry, which summarised findings on clusters and identifies critical and additional success factors in developing the cluster.
- The model "The NRC Cluster Framework" dealing with performance of clusters, presented by Canada's National Research Council (see Cassidy et al. 2005),
- The project "Cluster Benchmarking in Poland" implemented by the Polish Agency for Enterprise Development in co-operation with Deloitte Business Consulting S.A. in 2010, in order to analyse the cluster landscape and provide information for clusters on how to improve their performance, as well as to show cluster best practices thus facilitating learning processes amongst them (see PARP 2010 and 2012).

From these facts it is clear that an entire series of methods exist that deal with assessing the identified areas. In many of them, **benchmarking** appears to be a suitable additional tool both in terms of comparing performance and in terms of study, adoption and adaptation of good or best practices. However, so far none of the methods (models) of benchmarking assessment have been accepted for becoming (inter)nationally recognised within the framework of clusters and their members. A model using the possibilities arising from engaging into activities of clusters for assessing performance would benefit the

---

[1] Institute for Strategy and Competitiveness. 2003, "The Cluster Mapping Project". http://data.isc.hbs.edu isc/index.jsp (accessed October 20, 2008).

[2] Institute for Strategy and Competitiveness. 2002, "The Cluster Meta-Study", http://data.isc.hbs.edu/ MetaStudy2002Bib.pdf (accessed October 20, 2008).

[3] "A Practical Guide to Cluster Development. A Report to the Department of Trade and Industry and the English RDAs by Ecotec Research & Consulting", England's Regional Development Agencies, 2004, http://www.dti.gov.uk/files/file14008..pdf (accessed May 20, 2010).

management of companies when measuring the performance of searching for potentials in market competition.

Analysis of current practice using benchmarking procedures amongst clusters in the Czech Republic and abroad

For designing an organisation's own benchmarking model, it is essential to perform an analysis of the current state of applying benchmarking methods amongst companies and clusters here and abroad. Findings regarding the current approach of the management of clusters in assessing data on the performance of members of the cluster and the cluster as a whole, including use of benchmarking, were ascertained in the form of a questionnaire examination performed within the previously mentioned research projects.

As literature research has shown, benchmarking may be a handy tool for resolving many questions in the area of the performance of clusters and their members. Questionnaire research has answered the question of whether clusters offer this support activity to its members (Figure 4-1). The results show that roughly 50% of clusters use or plan to use benchmarking. Today, benchmarking is most frequently used in Austria, where half of its clusters apply benchmarking. There is a significantly lower proportion amongst Czech, Slovak and Polish clusters. Here, benchmarking is implemented in nearly 20% of clusters, but a series of clusters plan to employ benchmarking in the future. In comparison with the status in developed countries, the level of usage of benchmarking is still low; for clusters in modern countries, benchmarking is used in over 60% of them with another nearly 20% planning to use it.

The research also included ascertaining opinions of managers on factors that significantly influence the successful development of clusters. One of the factors was the use of the benchmarking method. The results show that despite the fact that benchmarking is not amongst the critical factors of successful development of the cluster, this factor in comparison with others is considered to be of average importance (Figure 4-2).

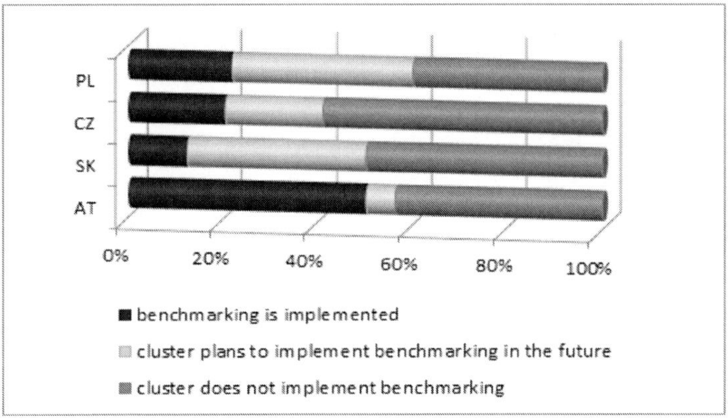

Fig. 4-1 The level of implementation of benchmarking activity according to the sample group of clusters in the selected countries
Source: Own development

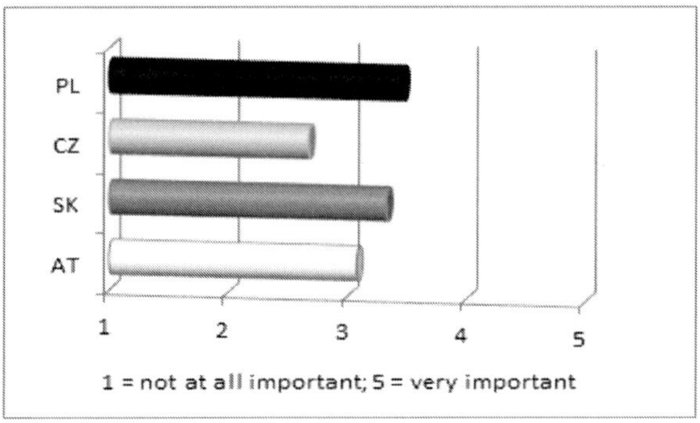

Fig. 4-2 The importance of benchmarking for the successful development of the cluster
Source: Own development

Research has further dealt with the question of whether clusters analyse their performance and that of their members. The research revealed an important difference between developed and more experienced clusters in Austria and clusters in the Czech Republic, Poland and the Slovak Republic, which are still in the early stages of development. Over two-thirds of Austrian clusters gather data about the performance of their members, but in the rest of the analysed countries this proportion is less than one-third, for example, in Poland it is a mere 10%. A similar trend appears in analysing the performance of the cluster itself (Table 4-3). In Austria, the vast majority (92%) analyse their own performance, ordinarily by using their own performance metrics (85%). Around half of clusters in the Czech Republic analyse their performance; some use their own performance metrics, whereas others use metrics determined by agencies providing support to clusters. In Poland and the Slovak Republic, the proportion of these clusters is significantly lower (22% in Poland, 12% in the Slovak Republic).

Table 4-3 Opinion of cluster managers concerning the analysis of cluster's own performance

Does the cluster analyse its own performance?	PL	CZ	SK	AT
Yes, using its own performance criteria	20%	25%	13%	85%
Yes, using criteria of the governmental/regional agency providing the financial support to the cluster	6%	25%	0%	38%
No	78%	53%	88%	8%

Source: Own development
Note: The possibility of multiple choice.

This shows that benchmarking has been implemented not only for assessing performance of individual companies, but also for clusters as a whole.

Benchmarking model of performance assessment of enterprises in clusters

This section presents possibilities of applying benchmarking to assess the performance of enterprises beneficial in this very cluster environment. The average performance of the cluster as a whole or only its strongest member may be a suitable benchmark for assessing a company's overall performance and for determining its position within the cluster. Meanwhile, the compared enterprise may be either a current member of the cluster or an entity standing outside of it, for example, a company considering joining the cluster and its results may be important for cluster management to deciding whether or not to accept the company into the cluster.

As critical literature research and compiled research projects implemented both here and abroad have shown, an appropriate combination of indicators is one of the most widely accepted prerequisites for an optimum performance measuring system. In the light of vast possibilities offered by clusters, its member may focus not only on "normative (metric) benchmarking", but also on the "process benchmarking". Traditional focus on financial measures is often criticised (for example Davies and Kochbar 2002) due to its inherent historical character and insufficient interconnection with broader performance measurements. Another criticism forcefully suggests that financial measures, as promoters of short-term thinking, serve as barriers to strategic thinking and innovations (Kaplan 1983). Gomes and Yasin (2011) emphasise that utility of non-financial measures and a balance and integrated approach are endorsed.

For example, the concept **Balanced Scorecard (BSC)** enables a complex view of the performance of the company with the help of the above-mentioned indicators. It augments financial metrics of past performance (delayed indicators) by including a new metric of moving forces of future performance (advanced indicators) and deals with relationships of causes and consequences between them by means of a strategic map.

These strong points of the BSC method, which convincingly fulfils criteria placed upon an effective concept of performance measurement and is thus one of the most widely applied methods of corporate practice, may be used to create a model. The benefit of using the BSC concept is assuring a balance between financial and non-financial or delayed or advanced indicators, which are designed in the model for assessing the performance of companies, including validating the idea of a map interconnecting four perspectives with the help of causal relationships amongst individual areas of performance.

If key performance factors are determined, including cause-consequence relationships between them, providing key performance indicators is easier because they are derived from just these factors. These indicators and their values declare in clearly quantifiable measures the level of company performance. Mainly beneficial was the monitoring of values of indicators over time or their comparison with other entities (other companies in the clusters). The existence of the cluster brings an indisputable advantage in the form of a relatively consistent sample for repetitive comparison.

Specific demands are placed within the benchmarking approach on input data. Such performance indicators must be selected that come from **objectively acquired input data**. This means that this data is not founded upon subjective opinions of the submitters, but on objectively verifiable facts whose source is mostly a company information system. A prerequisite for successful performance of intercompany benchmarking is also the **comparability of input data**. This means that all users understand the given term the same way, and a value is assigned to the selected input data that expresses the same thing in varying companies. A determined measuring unit serves this purpose, but so do formed descriptions of selected input data, mainly such data that could be understood in different ways.

Similarly to the majority of enterprise performance assessment systems, a large part of input data comes from accounting materials such as the balance sheet or profit and loss report, or other financial and economic information. It is also possible to use information from the area of purchase and sales, manufacturing, HR and innovation. For comparing the performance of companies, it is not appropriate to use indicators in absolute expression. These parameters are not suited to intercompany comparison, because their value within the given range is significantly influenced by a specific product, company manufacturing means, etc. For this reason, **relative (proportional) metrics** are applied that are relevant for comparing companies in the given field. In relation to the selection of appropriate performance indicators, emphasis must also be placed on **objectivity** and **minimum demands on time and money** in relation to obtaining input data.

To improve the quality of benchmarking, it is possible to apply other selection criteria, enabling comparisons with a sample of companies, which, in their field, cluster membership, locality, size or revenue or number of employees, most closely match the compared company. Only results of such a comparison may lead companies to finding justified deviations from performance of the benchmarking sample, and to the consequent adoption of appropriate measures leading to improving a company's own performance.

Outputs of the benchmarking model (for example a software solution) may be divided into several integral blocks. If benchmarking is built on the concept of a Balanced Scorecard, these blocks represent assessment of company performance within the framework of a financial, customer, process and growth perspective. The proposed model can be used as a web application offered to users from the web server over the Internet. It has to fulfil the requirements for user-friendliness, accessibility and dependability, and to represent low administrative and time demands on the part of the user and model administrator; at the same time, the software solution has to minimise the risk of unauthorised use of sensitive data by competitors. The goal of the created model that is designated primarily for the management of individual companies and clusters is a systematic assessment of current as well as a future potential level of performance of the organisation, and identification of weak points against the chosen benchmark such as average performance of the industry/field, members of the chosen cluster or of the best companies in the market, in the framework of assigned criteria (field, cluster, number of employees, size of returns, region). The model does not have ambitions to assemble an action plan in the shape of specific steps leading to an improvement of a current situation. The outputs of the model will serve as the basis for the acceptance and implementation of corrective measures with the goal of eliminating ascertained weak points and subsequently increasing the overall performance and competitiveness of the company in the market.

This section introduced an approach to using benchmarking for assessing the performance of enterprises within the cluster environment. Benchmarking enables a comparison of the achieved performance, identification of weak points in a company's performance, and adoption of suitable measures to eliminate these weak points, which may help improve the overall performance and competitiveness of companies. The average performance of the cluster as a whole or only its strongest member may be a suitable benchmark for assessing a company's overall performance and for determining its position within the cluster.

Cluster management teams may also use the presented performance assessment model when applying benchmarking in order to assess the performance of individual members of the cluster, to monitor development in the cluster's performance as a whole, or to compare the cluster's performance to development in the given field or branch. This approach and its outputs may be used by cluster management for deciding on whether or not to accept a potential party interested in joining the cluster. Using benchmarking in this way has thus become an invaluable source of data expressing the performance of individual companies, clusters and entire fields or branches, which may not be readily available.

A framework for successful cluster development in selected Central European countries

The most significant factors for creating an ideal environment for clustering were identified and a framework for cluster development was formulated based on extensive research of primary (interviews and questionnaires) and secondary data sources (critical literature research).

All the factors influencing cluster development can be divided into the macro (government, policy) and micro (cluster) perspective. There are also factors linking both perspectives together such as the question of financing, Triple Helix linkages, information flow and positive business environment creation.

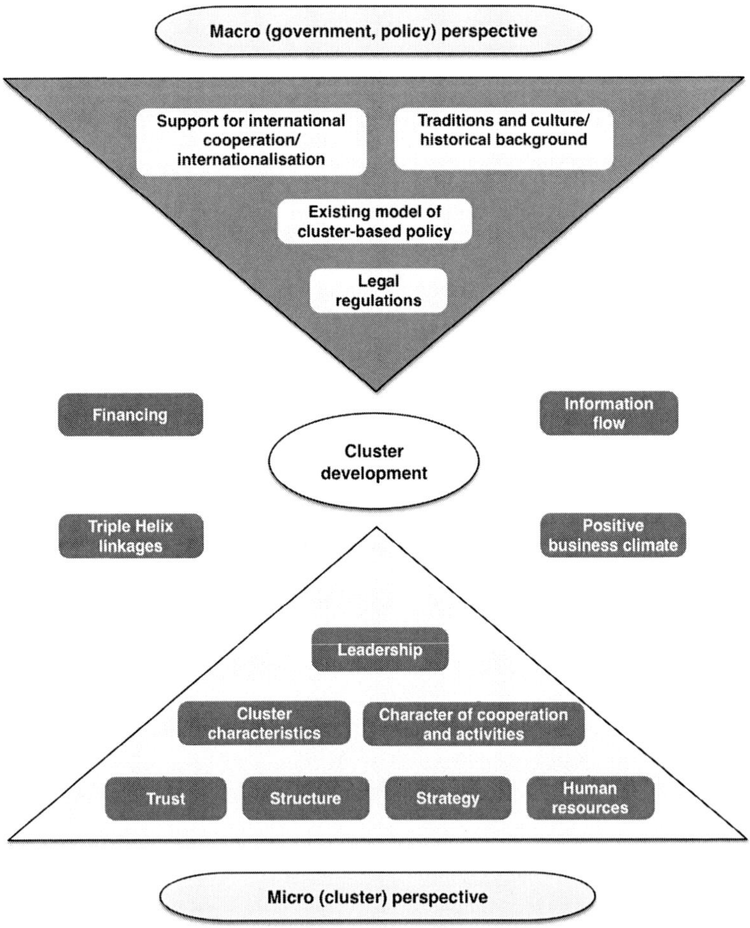

Fig. 4-3 Framework for cluster development
Source: Own development

There are also factors which are important for economic development in general such as macro-level foundations and adequate infrastructure (physical, institutional and legal), macroeconomic stability and availability of resources (typical for a given location), well-functioning markets (goods and services market, labour market, financial market, etc.) and the educational system. The existence of specified standards, the stability and persistence of economic law and an appropriate resource base in the country/region in which the cluster operates are fundamental issues for growth of co-operation within the cluster companies and the cluster structure itself.

Figure 4-3 presents a framework for cluster development. The conducive nature of the distinguished factors creates an ideal environment for the development of the cluster (organisation) and/or cluster initiative.

The following can be distinguished within the **macro factors** influencing cluster development:

a) **The existing model of a cluster-based policy:**

The existing model of a cluster-based policy appears as one of the most crucial factors conditioning the successful development of the cluster concept in a particular country. The model should therefore be based on a non-isolated and independent cluster policy characterised by transparent rules and regulations. In addition, this policy must represent a consistent system approach with clearly defined short, medium and long-term goals. In the case of European Union countries, the cluster policy activities on a country level must always be coherent with the one on the EU level.

The cluster policy must include a well-prepared strategy for cluster support with clearly specified strategic activities both on a national and regional level, as well as a focus on an analysis of existing conditions (business environment) and considerations as to the actual needs of the clusters and their views in designing changes in policy activity. It must also be focused on identification of leading clusters for regional and national economic development and their prioritised treatment with attention on, however, finding a balance between the prioritisation of a prosperous cluster (organisations)/cluster initiatives and a neutral approach (maintaining equal chances). Moreover, a vital element of a cluster policy consists of a balanced horizontality and complexity against the decentralisation of responsibilities and activities on the level closest to the given cluster (preferably the regional level).

Furthermore, a policy based on clusters should create new values as well as define and communicate a clear economic vision to all the actors in the economy.

An effective and consistent implementation as well as its respective "advertisement" constitutes inseparable elements.

b) **Legal regulations:**

Defined legal regulations are inseparably connected with an appropriate cluster policy approach. Clearly defined, coherent rules and regulations for cluster support, transparent legal provisions (legislation related to public assistance), and simplification of the administrative procedures (liquidation of bureaucracy barriers) condition a favourable environment for cluster concept development.

c) **Support for international co-operation/internationalisation:**

In light of the fact that international co-operation and the development of internationalisation processes amount to a necessity in maintaining competitiveness in today's global market, the promotion, initiation and facilitation of international linkages/co-operation appear as crucial. It is therefore vital that the cluster policy includes trans-regional and trans-national issues in its approach. In addition, financial support for international co-operation activities, their monitoring and evaluation should be incorporated.

d) **Traditions and culture, historical background:**

Another crucial factor consists of the positive aspects in a particular country or region such as, for example, long traditions of co-operation and/or conducive historical conditions (for example traditions in a particular industry area). When made use of and with simultaneous overcoming of potential cultural and language barriers, clustering has a flourishing perspective for its development.

The **micro factors** influencing optimal cluster development have the following characteristics:

a) **Leadership:**

Leadership ranks among the most crucial micro factors influencing successful cluster development. One of the most important characteristics of the leader (facilitator) is being neutral, charismatic, and characterised by credibility and integrity; having an intimate knowledge of the industry (experience) and relevant partners (a good network of contacts), as well as sufficient knowledge about all the cluster members; a person with appropriate leadership skills and status (a neutral communicator and listener), who is able to gain political and international access. A steering group-a supporting body, for example a cluster presidium, board, or council, should support the leader.

The position of cluster leader should be a full-time job with an attractive (i.e. rewarded with an appropriate salary) remuneration for being the cluster manager/co-ordinator.

b) **Cluster characteristics:**

One of the fundamental characteristics of the cluster is the presence of functional networks and partnerships within its structure. Additionally, the proximity of strong competitive behaviour constitutes an integral part of the cluster which can be characterised by a certain history, local dynamism, a degree of competition, the level of technology, the degree of competitiveness, global attractiveness and the global market reach.

Another crucial factor is the fact that the cluster should not operate as a subsidy-driven apparatus or as a result of a political decision but as a business-driven mechanism.

c) **Character of co-operation and activities:**

Inseparably connected with the aforementioned factor is the character of co-operation and cluster activities. Here attention has been devoted to stable co-operation amongst participants and their strong internal links. Active co-operation of cluster members who co-operate based on a common project implementation where a focus on teamwork is emphasised consists of an integral part. Continuous contact with potential cluster participants and the open character of the network also appear to be important factors. In addition, the cluster (organisation) should represent the ability to adapt to changes as well as to adjust the activities to the specifics of the region, sector and phase of cluster development. Furthermore, vital in their importance are the use of the elements of proximity as well as the identification of key sectors of activities where a focus on action, training and education schemes have to be distinguished. Additionally, clusters need to learn not only from their own but also from other cluster's experiences (best practices usage) and go beyond stereotypes in order to develop successfully.

d) **Trust:**

Trust constitutes a fundamental factor conditioning the existence of every relationship regardless of whether it is a business or personal relationship. Mutual trust amongst the partners in the cluster has therefore always been highlighted as the core of successful cluster development. The existence of an ethical codex inside the cluster network is also related.

With regard to the cluster concept, trust in government initiatives (for example organisation of events promoting clustering, implementation of programmes offering entrepreneurs and/or clusters oriented support) also appears as a significant component.

e) **Structure:**

Another important factor consists of the right form of cluster emergence, which has to be in compliance with the existing conditions in the country (i.e. bottom-up, top-down or mixed approach), and its legal form (formalisation of the cluster). Furthermore, a strong internal structure, separation of the thematic groups/industry groups as well as carefully considered cluster boundaries (flexible as the cluster evolves) influence successful cluster development.

f) **Strategy:**

Successful clusters are characterised by a clear vision and thinking over a long-term perspective. They have a defined strategy of activity (i.e. cluster niche, outlined activity, expected results, identified resources needed, identified priorities, milestones, a realistic time frame set up, etc.) and a formalised strategic approach to planned events. Furthermore, they indicate the optimisation of the management process, flexibility and a quick reaction to changes as well as a rapid response to changing customer needs. In addition, the creation of internal mechanisms for financing cluster activities appears as crucial for successful development.

g) **Human resources:**

The people who create the social network within the cluster consist of an integral part. Their passion, commitment, sufficient awareness as well as understanding of the clustering idea and its benefits condition the cluster's success.

The following can be highlighted as being among the crucial factors linking both the macro and micro perspectives:

a) **Financing:**

Financing, in the broad sense of the term, appears as one of the fundamental factors conditioning successful cluster development. Without sufficient financing no business structure can attain excellence.

Attention has been paid to the fact that clusters need to be self-sustainable and able to self-finance their activities and financial support and access to finance for clusters (organisations)/cluster initiatives are significant in their initial stage of development. This support, however, has to be based on adequate requirements defined in the programmes offering financial support and solid guidelines for applicants. Here, adequate availability of specialised and easily accessible tools supporting clusters, for example, informative portals (containing important and practical information) plays an important role. In addition, the documents of the programmes need to be simplified and written in an understandable legal language. Of particular significance is the possibility to apply for financial support from the government for cluster mapping and/or cluster development (the possibility to apply for support within an extended geographical area, for example, not only within one region). In addition, prioritised financial support for existing clusters should be balanced against support for new possibly prosperous initiatives. The specific character of the cluster in the project selection criteria (public support) should also be taken into consideration.

b) **Triple Helix linkages (public – private partnership):**

The links between entrepreneurship, education and research, and the governmental spheres are one of the main characteristics of the cluster, while harmonious and creative co-operation between industry, science and government authorities appears as crucial for the cluster's success. In this Triple Helix link the appropriate roles in the public - private partnership (public sector - as a catalyst, private sector - as a leader) as well as a solid private-public dialogue (between clusters and authorities) are vital to maintain. In addition, the adaptation of activities to economic reality (both by the authorities and the clusters themselves), collaboration on multiple levels and the connected consensus on key issues have to be present. Moreover, various fields of co-operation with public authorities have to be formalised.

The vital aspects of cluster development constitute the support at the regional/local level from regional/local authorities. This includes the engagement of the public sector (informational and financial support) in the early phase of the cluster development with a decrease in time, as well as a balanced level of intervention by authorities in general (both financial and organisational; balance between top-down and bottom-up approach). In addition, clustering processes may be influenced by the mutual co-operation of authorities (i.e. policy creators and policy contractors, co-operation between public administration, politics and economy, etc.). An appropriate knowledge and qualifications of authorities dealing with the cluster topic (for example offering consultancy) benefits the quality of the provided support. Additionally, the existence of an institution representing the cluster's interest and community wide involvement in the process of building new links between communities, business and government also appears as significant.

Effective and active co-operation with educational and R&D institutions is an inseparable element of the Triple Helix link.

c) **Information flow:**

Sufficient awareness and understanding of cluster participants regarding the clustering idea and its benefits (both by the business-entrepreneurs and government-authority spheres) as well as dissemination/popularisation of knowledge on clusters play a crucial role in cluster concept development. Additionally, the existence of a platform for information exchange (exchange both inside and outside the cluster) ensures and stimulates the flow of tacit knowledge.

d) **Positive business climate:**

The assurance of a positive business climate providing appropriate conditions for network co-operation, supporting collective research and innovation as well as entrepreneurship and providing conditions for the creation and development of new businesses including the development of local incubation centres, constitute inseparable elements creating a favourable environment for the development of clustering processes. The promotion of a culture of enterprise co-operation and networks as well as the creation of a mechanism for collective learning is an integral part of that. Other aspects such as the maintenance of a strong entrepreneurial culture and the support of competitiveness of companies via technology development also have a stimulating effect. In addition, an appropriate understanding of clusters as an economic phenomenon (an economic system based on market mechanisms not as a political or promotional tool) and the perception of company-cluster members as players in the developmental process contribute to a positive business climate creation.

The presence of distinguished factors creates an ideal environment for cluster development in a country.

On the basis of a comprehensive analysis of the aforementioned factors in the selected countries i.e. Austria, the Czech Republic, Poland and the Slovak Republic, the following elements of a framework (in an ideal environment) can be distinguished as the most significant factors determining the cluster concept development in these countries (Figure 4-4):

- From the macro (government, policy) perspective: **an existing model of a cluster -based policy and legal regulations**,
- From the micro (cluster) perspective: **leadership, the character of co-operation and activities, trust and structure**, and
- From both perspectives: **financing**.

The highlighted elements constitute core factors for cluster development in the analysed countries where other elements acknowledged within the framework proposal are lower in importance.

Simultaneously, the provision of these specified factors on an appropriate level in the selected countries constitutes a source of problems.

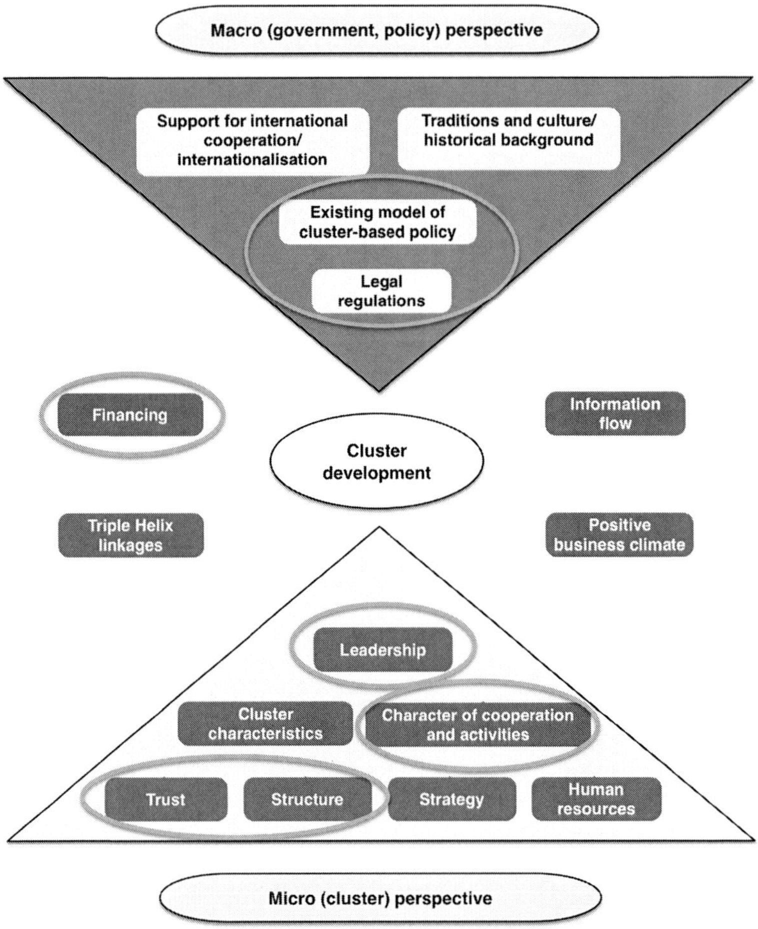

Fig. 4-4 Framework for cluster development in the selected countries with the most significant factors emphasised
Source: Own development

First and most important for all the selected countries, is to ensure the **existence of a model of a cluster-based policy** - independent and with clear rules and guidelines. All these countries, with the exception of the Slovak Republic, have already created the **support mechanisms** for cluster development providing support within various governmental as well as European Union programmes. Having said that the cluster policy does not appear to be apparent and explicit but as in Poland or Austria constitutes part of a usually complex system of policies such as regional development policy, innovation or technology policy, etc. Additionally, in Austria several strategic national guidelines implicitly focus on cluster policy, and guidelines for regional development allude to cluster development (here, the type of cluster policy can be described as a policy of national frameworks for the regional cluster policies, with different intensity and steps undertaken). In contrast, in the Czech Republic, programmes on a national level are explicitly focused on clusters and the National Cluster Strategy was implemented providing strong support for the cluster concept development. The initial strategy, however well launched, did not meet expectations, was consequently neglected and left the country without an explicit cluster policy. The need to revive and implement the model of a cluster-based policy is therefore being emphasised.

The introduction of a comprehensive way of thinking in connection with a cluster-based policy (i.e. appropriate strategy and government policy) is recommended.

A cluster-based policy should incorporate a diversified level of development existing in particular regions in the country (for example, in Austria particular regions indicate a different level of activity supporting clusters) and the government activities should be intensified to a level closest to the clusters (i.e. local and/or regional).

Effective monitoring methods allowing for an estimation of the results of the activities already undertaken need to be developed and introduced. The recognition of cluster-based policy effects would allow government authorities to become engaged more effectively in the future process of cluster creation and development and proceed with adequate steps towards cluster development in further programming periods.

Inseparably connected with the appropriate cluster-based policy implementation is the **existence of adequate legal regulations.** Government regulations should be formulated in a way so as to not hamper and restrict cluster competitiveness (the elimination of problems connected with the use of certain sources of support by cluster members). Rules and regulations should be clear, transparent and consistent.

It is initially recommended to establish a legal title for the cluster (as a cluster is a specific beneficiary) as cluster structures are often regulated (take the form) in a manner appropriate for other entities, for example, associations, foundations, which can twist the cluster idea. Additionally, it is advisable to enforce tools determining clear ownership within the cluster structure.

From the micro (cluster) perspective, the analysed countries need to focus attention on the **leadership** component.

Companies associated within the cluster have to have a common vision of functioning as a network, only then they can take active participation in a cluster (otherwise the cluster remains in the latent form). A significant role is therefore played by the properly selected leader of the cluster organisation who should have the required knowledge about the area in which the cluster operates as well as about all of its members, and who should formulate a long-term transparent vision of the cluster organisation in accordance with all its participants. The leader has to co-ordinate the activities of all the cluster members and stimulates the exchange of information and knowledge among the cluster participants (the use of synergy potential). The leader may be the cluster manager, a key large (dominant) company in the region, or even a business agency (for example EcoPlus in Lower Austria).

Unfortunately, one of the major problems in regard to leadership is that clusters are often afflicted with understaffed management and/or lack of a manager/co-ordinator fully engaged in the cluster activity (for example, in Poland and the Slovak Republic many cluster managers work on a voluntary basis having another primary job). The elimination of weakness regarding a lack of human resources in the cluster as well as ensuring funding sources for the professionalisation of the cluster – the establishment of cluster headquarters (office), the formalisation of its structure and activities, covering the costs of the co-ordinator activity, etc. are therefore necessary.

Further attention should be paid to the **character of co-operation and activities** and the inseparably connected components of **trust** and **structure**.

The clusters in the selected countries are confronted with the weakness of character of internal co-operation often caused by incorrect apprehension of the clustering idea (the treatment of participation in the cluster as a secondary thing) with the trust issues resulting in limited openness to co-operation. Clusters often mention internal deficiencies caused by the conservative attitude of the leadership and the inability of companies to co-operate in a team as being among the most serious problems encountered. A lack or limited trust among cluster participants and thus their apathy and/or passiveness as regards cluster activity are the results of misunderstanding of the clustering idea. Many companies are not willing to co-operate as they "were not taught" how to co-operate in a team (lacking the tradition of business-to-business co-operation). In addition, a mental barrier, often ignorance and/or simply parochial thinking and fear of being abused by other members (competitors) are present.

It has been observed that the clusters in all the selected countries focus their priority on networking activities with the aim of stimulating development and strengthening co-operation networks and thus the integration of the environment. Managers' point out that those activities rank among those of high interest amongst cluster members. The creation of co-operation ties has been emphasised as one of the most important achievements accomplished by clusters, which confirms the difficulty of building those ties. Clusters underline an understanding of the clustering idea and the creation of co-operation linkages among members as problematic.

The clusters in the selected countries encounter numerous difficulties due to distrust, inability to find a consensus and the high workload of members. SMEs often do not see the value and do not use the opportunities to co-operate with other companies (lack of knowledge regarding the sense, the aids and benefits of establishing and functioning within the cluster organisation).

Therefore, as a crucial step, it is recommended that trust building be stimulated (increase the propensity of entrepreneurs to co-operate with other companies from the same or a related branch, as well as with other clusters), stimulate the spirit of co-operation among companies (their readiness and willingness to co-operate, their active participation in cluster activity, finding a common consensus among members) and enhance better time management on the part of cluster members (improve their regularity).

Additionally, there should be a focus on overcoming and eliminating the conservative attitude of professional executives and staff, through an increase of their competences, and strengthening the substantive of leaders (ensuring adequate training of the leaders, cluster facilitators). A focus should also be placed on the development of informal and lax contacts among cluster participants thus stimulating the trust building process; organisation of meetings where members can present their suggestions, exchange ideas and discuss common interests; the creation of a cluster database available to all members (thus influencing a higher level of co-operation), etc. In addition, increases in involvement of additional members, elimination of shared personal conflicts with members who are already in the cluster (a common barrier among the Slovak clusters), are recommended. Further attention concerning the stimulation of the development of trust and co-operation links between entrepreneurs and the academic sector (especially in Poland and the Slovak Republic) is compulsory. Mentality issues in regard to the university-business relationship and overcoming the institutional rigidity of universities and research institutes need to be eliminated - improve the information flow and communication between them, build and strengthen connections between the business sphere and R&D and educational institutions (ensure a common understanding, consensus, development of ideas for co-operation, building up a sense of security and satisfaction of scientists, etc.). The engagement of research and development institutions from the beginning of the cluster functioning appears as an important element as well.

The most significant factor influencing the successful development of clusters from both perspectives is **financing**.

Clusters in the selected countries face problems with financing, where the financial support for clusters is diverse (in regions), inadequate or is simply missing. In Austria, particular regions indicate different levels of intensity of activities with cluster-oriented support. In Poland, it is difficult for clusters to map out all the possibilities for applying for financial support (the sources of support are dispersed within different programmes) and there are also often changes in rules and requirements set within the supporting programmes which cause unnecessary confusion and discourage applying for the support provided. In the Czech Republic, the bureaucracy and organisation problems on the side of the implementing agency during the OPIE support programme has resulted in losses or poor faith in cluster-based policy functioning in the country. In addition, an important type of support offered for the cluster mapping and establishment phases was removed from the currently implemented OPEI programme meeting with great disapproval. In the Slovak Republic, marginal treatment of cluster issues and therefore no financial support from the government has resulted in missing the potential of clusters.

It is therefore recommended that adequate financial support be ensured from the government on a national level (respectively on regional levels) for start-up of cluster activities (for new potentially prosperous initiatives) as well as support for cluster functioning. In addition, appropriate tailor-made support for the development of already existing networks of collaboration with potential should also be provided.

Since clusters are characterised by a short period of existence and difficulties in tightening co-operation links based on trust due to mentality issues, they face the burden of unwillingness of businesses to financially contribute to cluster activities without seeing tangible benefits (declined sponsorship from companies). Without their financial engagement, the further development of cluster organisations is hampered and this is where the vicious circle closes. Clusters in their first phase of development have limited resources for financing the employment of a cluster animator (facilitator), management and/or the cluster's own employees, for learning from other experienced clusters (for example study visits for employees, mutual learning with other businesses), and for intensification and enlargement of activities or the resources necessary for applying for financial support for the implementation of a project. Financial support is vital in this stage of development where co-operation links and trust building takes place.

Attention should therefore be directed at financial support for the development of cluster excellence first through the professionalisation of clusters, strengthening the internal structure, and financing current operations at the first stage of cluster development (the Upper Austria activities in this regard should be followed). Additionally, in the Slovak Republic, a special focus on financial support should be placed in an even earlier phase of implementation of the feasibility study. Attention should also be paid to supporting the economic activity of new companies in the cluster (access to loans for SMEs). Once the cluster establishes a strong internal structure, further attention should be paid to its innovative development and development of internationalisation (stimulating the development of international co-operation links of clusters and companies collaborating within their structures; creating links and connections of regional clusters in the EU). Here, financial support for projects focused on R&D and/or inter-cluster collaboration development (also on an international level) should be in accordance with the market character of the cluster.

It is vital to ensure the availability of adequate financial instruments, which should be prepared on the basis of an analysis of the existing demand (attention to cluster needs and their stage of development and thus better targeting of existing instruments) and thus constitute help in fulfilling the desirable objectives within the support programmes.

Additionally, in Poland it is recommended that financial support resources be obtained, in order to facilitate and streamline potential beneficiaries to apply for the offered support, which are dispersed throughout numerous programmes on the national and regional levels. Furthermore, both in Poland and the Czech Republic, problems with grant settlement need to be eliminated (too detailed requirements) through simplification of the evaluation methods, as well as through creating appropriate support instruments on a regional level (which is currently characterised by poor or lack of financial support for clusters).

In addition, it is important to offer the possibility to apply for financial support for individual institutions creating platforms for cluster co-operation.

CHAPTER FIVE

THE NEW CONCEPT OF INTER-CLUSTER CO-OPERATION

The rapid development of globalisation and internationalisation has brought with it new perspectives on the issue of competition and co-operation as well as on networking processes including clustering. Clusters are being perceived in a broader context, where many of their attributes "notably their emphasis on networking and proximity based interaction – are now feasible on a global scale" and where "firms are paying much less attention to location within one country or another" (Innovating Regions in Europe n.d., Szanyi et al. 2007). The aspects of collaboration have been broadened from those within a cluster on a regional, national or international level, into a new type of inter-cluster co-operation in a worldwide context (Bathelt, Malmberg and Maskell 2004). Clusters began to strive to achieve the status of World-Class Clusters (WCC). These, according to the "White Paper"[1], rank amongst the high performing clusters, which can be characterised by 15 criteria divided into three categories: framework conditions, competitiveness of main cluster actors and high quality of cluster organisation/management. Amongst others, internationalisation appears as one of the important criteria influencing cluster development. The aspiration of being internationally visible, with co-operation links across borders is vital in this era, when clusters are no longer perceived as isolated organisations but as entire eco-systems constituting a new tool for competitive strategy, stimulating innovation, growth and competitiveness on a wider scale. Therefore, the potential for rapid development of innovations and strengthening competitive advantages has been noticed and a new approach of inter-cluster collaboration has begun to grow in popularity.

---

[1] Europa InterCluster. 2010, White Paper. The Emerging of European World-Class Clusters. Belgium, http://www.intercluster.eu/images/stories/white_paper/white_paper_the_emerging_of_european_world_class_clusters.pdf (accessed February 8, 2011).

The programme INTERREG IIIC and its project CLOE, which appeared as one of the few initiatives in Europe devoted to this issue, simultaneously providing a platform for clusters collaboration has brought attention to and promotion of cross-border, trans-national and inter-regional co-operation (Bialic-Davendra 2011).

In 2006, a new innovation strategy[2] was adopted by the EU highlighting the importance of strengthening clusters which influence the stimulation of promotion of innovation and the environment conducive to innovation development.

Two years later, the European Cluster Memorandum[3] was introduced focusing on promoting innovation in Europe through clusters, and pinpointing the need for stronger clusters based on innovation and excellence. Additionally, it underlined the need for active collaboration amongst European clusters in order to reach their full potential through creation of solid links to clusters and markets located elsewhere (Bialic-Davendra 2011).

Further, in 2008, the Brussels European Council stressed the importance of the development of broadly perceived innovation "through improved science-industry linkages and world-class innovation clusters and development of regional clusters and networks"[4].

Moreover, the European Commission launched a communiqué entitled "Towards world-class clusters in the European Union: Implementing the broad-based innovation strategy"[5] with an emphasis on the development of cluster excellence through intensification and deepening of trans-national cluster co-operation, and thus resulting in the creation of a European research and innovation space (Bialic-Davendra 2011).

In response to the European Union policy regarding clusters, V4 countries reached a "Memorandum of Understanding" in 2009. The parties committed to collaboration in the development of clusters and to

---

[2] European Commission. Putting knowledge into practice: A broad-based innovation strategy for the EU. COM(2006) 502 final. http://eur-lex.europa.eu/LexUriServ/site/en/com/2006/com2006_0502en01.pdf (accessed July 12, 2010).

[3] "European Cluster Memorandum", http://www.proinno-europe.eu/NWEV/uploaded_documents/European_Cluster_Memorandum.pdf (accessed December 7, 2010).

[4] Presidency Conclusions of the Brussels European Council (13-14 March 2008). http://www.consilium.europa.eu/ueDocs/cms_Data/docs/pressData/en/ec/99410.pdf (accessed December 8, 2010).

[5] European Commission. Commission Communication "Towards world-class clusters in the European Union: Implementing the broad-based innovation strategy". COM(2008) 352 final.

encouraging cross-border co-operation between them, in order to support co-operation both within the member states of the European Union and other countries.

The concept of inter-cluster co-operation amongst countries and world-class clusters began to increase in popularity and steps to "remove barriers to trans-national cluster co-operation, and to encourage the emergence and consolidation of world-class competitive clusters across Europe"[6] were undertaken. The co-operation linkages among clusters appeared to expand rapidly and take various forms such as: platforms for co-operation, clubs, alliances, trans-national co-operation, groups of collaboration, bilateral agreements, and partnerships, etc. (Bialic-Davendra 2011).

Amongst others, the European Innovation Platform for Clusters deserves attention. The platform focuses on cluster organisations from different countries aiming at promoting and facilitating active co-operation between them, across various member states and regions. While being open to co-operation with other initiatives, it provides an opportunity to associate with different partners with a focus on specific cluster sectors such as, for example, eco-innovative industries (EcoCluP cluster partnership) and biotechnology (ABCEurope cluster partnership). Another example of a world-wide, sector oriented network of clusters is the European Aerospace Cluster Partnership (EACP). This, while associating clusters from the aerospace industry, constitutes a platform supporting concrete projects and creating an environment for long-term trans-national co-operation among member clusters. An additional example consists of the Food Innovation Network Europe (FINE) strengthening co-operation within and between food clusters.

Additional steps stimulating inter-cluster co-operation include the enforcement of funding programmes for international and cross-border co-operation projects for clusters such as, for example, the European Territorial Co-operation Programme with its projects: ClusterCOOP (Central Europe Programme) supporting trans-national, inter-cluster co-operation in Central and Eastern Europe, Clusters CORD aimed at the creation of so-called "meta-clusters", CNCB, CluStrat, etc., or facilitation of trans-national networks between clusters from partner cities such as within the CLUSNET project.

---

[6] "Innovation Clusters in Europe: A statistical analysis and overview of current policy support". DG Enterprise and Industry Report, http://www.proinno-europe.eu/admin/uploaded_documents/innovation_clusters_in_europe.pdf (accessed February 27, 2011).

Another example consists of Netmatch (Innovation Network Denmark)[7], a unifying platform supporting innovation networks/clusters in Denmark, creating a platform for companies and research institutions to exchange knowledge and enter into collaboration projects, favouring competence development, driving an internationalisation agenda and thus providing a significant boost for Danish clusters and networks.

Attention to clusters has also been provided by the European Commission's proposal for a new "Europe 2020" strategy, in which cluster issues constitutes one of its flagship initiatives.

Furthermore, the European Commission[8] facilitates numerous initiatives connected with networking on different levels. Within PRO INNO Europe, it aims to integrate cluster policies focusing on stimulation of trans-national policy co-operation in regard to innovation. Through the Regions of Knowledge (RoK) Programme, it pursues trans-national networking of regional research-driven clusters and through Regions for Economic Change co-operation between regions. Networking between clusters across Europe stimulates the Europe INNOVA initiative. PRO INNO Europe and ERAWATCH constitute platforms for diffusion of the information about cluster programmes and priorities set up by the EU. Additionally, other instruments are being undertaken in order to stimulate trans-national co-operation between clusters.

As stated in the European Cluster Memorandum:

> Policies set at the European level have an important influence on clusters, especially their emergence and their transnational linkages. Also, clusters become more visible and attractive if they have strong linkages with related clusters in other regions and countries.

Despite the fact that a number of activities aimed at stimulation of inter-cluster collaboration across different regions and national borders in Europe have already been undertaken, the concept is developing slowly. The European policy towards clusters and support within existing cluster policies in specific countries appear to be unsystematic and difficult to implement in practice. As each member state has its own approach, various solutions are being implemented (Bialic-Davendra 2011, Europa InterCluster 2010). Therefore, the co-ordination of activities undertaken

---

[7] www.netmatch.nu (accessed May 5, 2011).

[8] "Innovation Clusters in Europe: A statistical analysis and overview of current policy support". DG Enterprise and Industry Report, http://www.proinno-europe.eu/admin/uploaded_documents/innovation_clusters_in_europe.pdf (accessed February 27, 2011).

across different countries requires improvement, and trans-national co-operation between clusters needs further strengthening.

Improved support for international cluster co-operation and its significance are highlighted by the ECPG in their policy recommendations[9]. The ECPG pinpoints that creating strong links between clusters in different locations is a way to achieve a level of world-class excellence and competitive advantage above other independent clusters, as well as stimulating specialisation and integration into European or global value chains. Therefore, the recommended steps focus on providing data for the actual cluster landscape, promoting strategic and commercial networking of clusters as well as the international mobility of their actors.

The emphasis on the importance of international collaboration between clusters as a way of stimulating and maintaining their competitiveness in the global market is being observed. Gorynia and Jankowska (2007) aptly state that the high competitiveness of clusters may constitute an incentive for international expansion of cluster activities, where internationalisation may provide a chance for increasing cluster competitiveness. Various benefits can be observed including benefits of scale resulting from the penetration of foreign markets or the need to increase the quality of products and/or services offered, etc. The development of cluster international co-operation may also enable access to specialised and advanced resources, know-how, a qualified work force, access to new customers and/or suppliers, to specialised services and infrastructure, to innovative branches in the industry in which the cluster operates and also proximity to competitors (what increases the pressure on innovations).

Clusters do realise the importance of development of inter-cluster co-operation both on the national as well as on the international level. Within the conducted research, the significance of this issue was examined among clusters from Austria, the Czech Republic, Poland and the Slovak Republic. Figure 5-1 presents opinions of managers from the sample group of clusters regarding the strategies dominating inter-cluster co-operation.

Establishing contacts with other networks and clusters, as well as participating in co-operation projects with their members appear to be leading strategies. Simultaneously, searching for potential partners for co-operation and involvement in shared project implementations are the first steps undertaken when starting inter-cluster co-operation. Among Polish clusters, concrete plans for entering eligible target markets as well as establishing contact with business development agencies are also popular.

---

9 European Cluster Policy Group. 2010, Consolidated Set of Policy Recommendations on Four Themes. http://www.proinno-europe.eu/ecpg/newsroom/ecpg-final-recommendations (accessed August 22, 2011).

Austrian clusters further focus their attention on entering eligible target markets and participation in business trips. In contrast, Czech and Slovak clusters appear to be further concentrated on the establishment of contacts with other networks and clusters by joining technology platforms, as well as contacts with business development agencies. Only a minority of (Polish and Czech) clusters claim that there is no strategy or concept available at all, which testifies to a strong awareness of the need to develop co-operation links not only inside but also outside the cluster organisation.

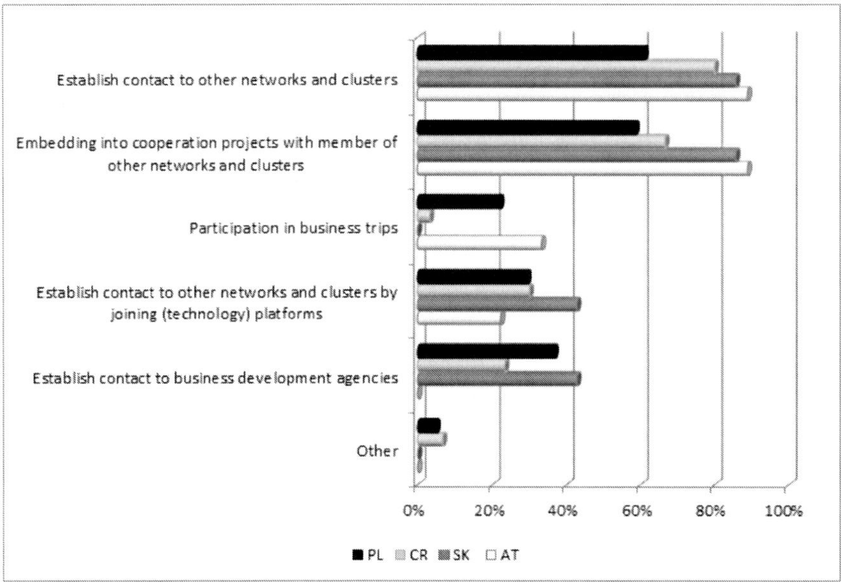

*Number of respondents according to the country of origin: PL – 41, CZ – 30, SK – 7, AT – 9

Fig. 5-1 Opinions regarding the strategies dominating inter-cluster co-operation according to the sample group of clusters from selected Central European countries
Source: Own development

Within the analysed sample, 24 (59%) of Polish clusters, 21 (68%) of Czech, 6 out of 7 Slovak, and 9 out of 10 Austrian clusters declare co-operation with at least one other cluster. The character of inter-cluster co-operation is often a mixture of the formal and informal form of collaboration depending on the implemented activities.

The most popular activities implemented within the informal type of co-operation, regardless of the country of origin of the analysed clusters, include the exchange of information and experience in cluster management, specific for the industry, and in the area of obtaining financial support, as well as networking. These activities often constitute the basis for the development of future formal co-operation links within, for example, specific projects focused on shared promotion, research, trade, etc. The organisation of various worldwide conferences, cluster manager meetings, match-making events, trade fairs and other events pose an excellent opportunity for networking and meeting potential partners for collaboration. The informal character of co-operation of Polish as well as Austrian clusters also predominates in project preparation (consultancy), shared research and development, marketing, collaboration in new cluster creation and benchmarking. Additionally, Austrian clusters informally implement activities in the area of the products or services offered. In the Slovak Republic, attention is also focused on the implementation of shared research and development under informal activities.

In contrast, the formal character of co-operation predominates in the organisation of seminars and conferences in all the selected countries. In Poland, this is additionally the case in education and training, as well as in joint participation in international events (trade fairs) and products/services offered. In the Czech Republic, in turn, this is also the case with shared project implementation. Among the Slovak clusters, project preparation (consultancy) and their implementation (for example EU projects), the organisation of shared events as well as joint participation in international events, common marketing, products/services offered, and benchmarking have a formal character. In terms of Austrian clusters, the formal character of activities predominates in the area of education and training, organisation of shared events as well as joint participation in international events, and shared project implementation.

The responsibility for inter-cluster activities among Polish clusters lies in the hands of the cluster manager (96%) with infrequent (29%) help from the top management of the companies in the cluster (management board). Similarly, in Czech clusters the responsibility belongs to the cluster managers (81%), however, sometimes with the support of the top management of the companies in the cluster/management board (29%) and/or representatives of a university or research institution (14%). Member companies and representatives of governmental/regional institutions are not involved at all. In the Slovak Republic, a cluster manager (in 100%), often with the help of the management board and/or companies themselves, supervise the implementation of inter-cluster activities. In Austria, a cluster manager is responsible for inter-cluster collaboration in 100%, rarely with the support of the advisory board which may consist of representatives from companies, university, research or government institutions.

The arguments for cluster internationalisation, and thus development and implementation of inter-cluster activities may vary individually for each cluster. Figure 5-2 depicts the reasons for inter-cluster co-operation highlighted by a sample group of clusters according to their country of origin.

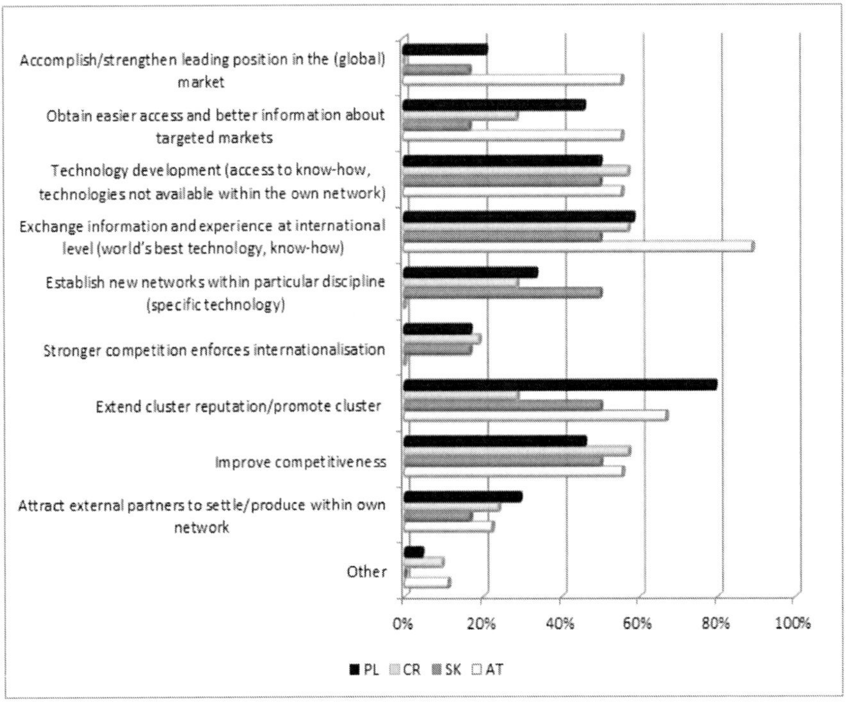

*Number of respondents according to the country of origin: PL – 24, CZ – 21, SK – 6, AT – 9

Fig. 5-2 Reasons for inter-cluster co-operation according to the sample group of clusters from selected Central European countries
Source: Own development

While the purposes of this type of collaboration differ amongst countries, one of the most popular appears to be the exchange of information and experience at an international level (world's best technology, know-how, etc.). As confirmed by the above, the exchange of knowledge connected with networking events facilitates the establishment of new contacts and is thus the first steps for the development of co-operation links amongst partners.

Major attention amongst Polish clusters is attached to activities promoting clusters and broadening their reputation for a wider group of potential customers or business partners. Austrian clusters view it as the second most important purpose of their inter-cluster co-operation.

All clusters, regardless of their national backgrounds, pinpoint technology development and improvement of their competitiveness as vital reasons for the development of co-operation with other clusters.

Additionally, Slovak clusters appear to place more attention on establishing new networks within particular disciplines (specific technology), while Austrian ones highlight the significance of accomplishing and/or strengthening the leading position in the (global) market. This in turn confirms the level and character of the development of clusters in these countries.

Despite the fact that the awareness of the need for development of inter-cluster co-operation is high, the current status of the research conducted confirms that not all clusters implement this type of co-operation. The reasons behind this may vary. A general barrier consists of the fact that a cluster is not prepared for an inter-cluster co-opetition because of its too young existence, undeveloped, weak internal structure, etc. An additional significant problem involves the financial barrier including a lack of funds in general, unsupported financing of common projects, problems with co-financing the projects, for example, with a cross-border character, connected with the fragmentation of cluster-based policies implemented on national, regional and local levels. At times lack of contacts/opportunities to begin co-operation appears as well.

Inter-cluster co-operation may result in both positive aspects stimulating cluster development as well as negative ones, hampering its activity. Among the positive features distinguished by the analysed group of clusters are:

- Exchange of information, experience and good practices;

Clusters value the opportunity for knowledge transfer between the co-operating partners the most. This type of co-operation creates favourable conditions for learning from one other about the way clusters function (also in other countries) and about existing cluster policies (around the world), new trends in global markets, new technologies, etc. The exchange of good practices on cluster management as well as other aspects of clusters development, experience with the use of financial sources (subsidies), constitute a significant advantage for inter-cluster co-operation.

- Broadening the network of contacts and partners;

Co-operation with other clusters favours the expansion of the base of contacts, obtaining new customers, entering new markets, broadening business horizons, obtaining international contacts with universities or research institutions, etc. Additionally, it creates the opportunity for the development of cross-company co-operation between cluster organisations or opens up the possibility to collaborate with entities with which no co-operation has been possible thus far.

- Joint promotion;

Through inter-cluster co-operation, partnering clusters may take advantage of a joint promotion, on the international level as well. In addition, new promotion possibilities may appear such as shared stands at international trade fairs, joint delegations, etc.

- Co-operation on shared (international) projects;

Clusters often collaborate with one another on shared project preparation (consultancy) and their implementation, for example, within 7FP, INTERREG, programmes for the development of a cross-border co-operation, providing advantages of new services and/or products offered, access to project funds amongst other aspects.

- Synergy of cluster resources;

Accumulation and synergy of resources and potential may stimulate improved and faster technological and innovative development of partners. Shared research and development may result in the implementation of new solutions in the innovation of products and/or processes, etc.

- Mutual business co-operation;

This brings the advantage of positive effects of economies of scale thanks to, for example, mutual purchases, the division of orders. In addition, joint marketing activities may result in an increase in awareness about the provided products or services, and broaden the spectrum of potential customers. Additionally, business co-operation with other clusters and their members may influence an increase in the business position of a partner cluster in the eyes of its business surroundings and thus stimulate an increase in its competitive advantage.

- Improved lobbying opportunities;

Often associating a large number of entities, cluster organisations through their inter-cluster co-operation and joint efforts have more power and greater opportunity for lobbying in favour of subsidies, more favourable legal regulations, a better infrastructure, etc. They have a "stronger voice" in communication with respective authorities and the shaping of a cluster-based policy in the country.

- Economic effects;

Economic effects result from the aforementioned synergy effects and mutual business co-operation. An increase in turnover, revenue, decrease in expenses on research, costs of organisation, projects preparation, etc. are often highlighted.

Apart from the aforementioned advantages, negative aspects (threats) resulting from inter-cluster co-operation can also be distinguished. Among them, the analysed clusters identify the following:

- Distrust;

One of the major threats within the members of co-operating clusters is distrust of new partners. Overrated personal contacts, not knowing the right people from other clusters, and thus difficulty in finding good business partners are also a frequent problem. Additionally, distrust is often connected with mental barriers which are often manifested through a lack of adequate openness and the conviction that everyone can benefit and that competitiveness stimulates further development. An equally frequent problem consists of a simple inability to co-operate. There may also be a lack of clear rules regarding the transfer of know-how and the related fear of transfer of knowledge and new technologies, fear of copying the ideas or intellectual property theft (for example patents), and thus the need to safeguard the know-how by cluster members. In addition, the possibility of migration of cluster partners to other cluster organisations appears as a further negative aspect.

- Passiveness;

Another negative aspect consists of reluctance and tardiness on the part of entrepreneurs to participate in activities within inter-cluster collaboration and their too weak commitment due to fear of the unknown. Therefore, the need for better understanding of cluster concepts, more cluster oriented thinking of their current and potential members, as well as the creation of the feeling of belonging to a cluster, are being observed.

- Rivalry;

Additionally, strong rivalry between the clusters and their members, and their fear of competition between the two sides poses a threat to this type of collaboration. Furthermore, clusters from one country and/or region may feel some level of competition when applying for (financial) support for their development. Another threat may constitute strong competition on the market and the related fear of losing the market position due to the acquisition of the local market by companies from the outside. As clusters their members co-operate and compete at the same time, a divergence of cluster objectives i.e. potential conflict of interest (everyone wants to promote their own products) may also arise.

- Lack of time or capacity;

The job of a cluster manager co-ordinating co-operation among numerous entities in a cluster is extremely difficult and time consuming (they often voluntarily lead the cluster, while being employed somewhere else), in many cases, there is not much time left for the manager to lead steps aimed at the development of co-operation with other clusters. Furthermore, as the number of cluster employees is usually limited to a cluster manager and very few additional employees engaged in its management (usually the project manager, financial manager, assistant and/or accountant), due to insufficient capacity the possibilities for internationalisation of cluster activities are restricted.

- Lack of financial resources;

Often lack of sufficient or limited available financial resources hampers the development of inter-cluster co-operation between clusters particularly from different countries. Here, the high costs of foreign travel, study visits, etc. and therefore difficulties in communication may discourage undertaking decisions regarding the development of co-operation with a particular cluster or simply force the postponement of these activities until the future.

- Problems in communication;

Communication problems may be connected with the aforementioned lack or limited financial resources for official visits to the targeted potential partner clusters or companies. They may also be caused by the language barrier, cultural differences, etc. Moreover, due to the fact that clusters function differently in various countries, differences in legislation may also appear, causing misunderstandings in the implementation of shared activities and unnatural elongation in decision-making processes (within a group of co-operating clusters).

- Misunderstanding the clustering idea;

Misunderstanding or abuse of the cluster concept for financial reasons (for example obtaining financial support in the form of subsidies from the government) may decrease the meaning of good cluster initiatives and discourage the development of co-operation links among potential partners.

Clusters are acknowledged as an important tool for internationalisation, stimulation of global economic growth and competitiveness. A crucial factor for their survival in a highly competitive environment consists of the development of inter-cluster co-operation links and facilitation of access on the part of SMEs to internationalisation through clusters. The outward linkages between companies inside and outside of the cluster are an important component for obtaining or maintaining the competitive advantage above others.

The basis for the development of inter-cluster co-operation on an international level requires the searching out of common interests in sectors, markets and technologies of the potential partners for collaboration and a strong focus on the part of the clusters on their objectives. International collaboration is not only conditioned by seeking for complementarity but also by stimulation of member companies in a cluster to provide the cluster managers with a mandate for internationalisation and to foster their mobility.

Additionally, often a lack of an existing strategy and concrete steps on the governmental level is not conducive to the development of inter-cluster collaboration. Therefore, a crucial point constitutes building awareness among policy makers to include internationalisation in the funding programmes and thus inspire others to create a collaborative environment and facilitate its development. In addition, Dervojeda (2011) has aptly noticed that "cluster managers all over the world operate in a turbulent, uncertain and highly complex environment" with the need for more synergies between policies at all levels, especially in European countries, being highlighted. Also, as clearly stated by Andersson et al. (2004: 149):

> A higher degree of collaboration and co-ordination of activities between European clusters could, if managed professionally, reap further potential positive effects and result in a higher innovative performance and growth in the long run.

Therefore, the creation of improved channels for the development of internationalisation and inter-cluster co-operation, as well as improved and more integrated cluster policies, where success can be clearly measured, is essential in order to achieve international prominence and strength through clusters.

CHAPTER SIX

CONCLUSION

A highly complex environment, continuous changes to economic or political conditions and rapid technological progress confront companies (businesses) with numerous challenges. Co-operation within a cluster appears as one possible answer to the increasingly growing competition and expanding globalisation processes. Collective initiatives by cluster actors allow them to utilise their joint potential and react faster and more effectively to global market changes.

However, clusters as a specific type of network are characterised by special needs and goals. An understanding of these needs, enablers and barriers, should transform the implementation of an appropriate policy and measures for support of clusters. Support mechanisms, clear rules and guidelines, as well as a felicitous strategy and government policy constitute one of the factors influencing cluster existence and development. Therefore, increased attention was devoted to the topic of cluster policy in this book with a special focus on their status in the selected countries.

Furthermore, successful co-operation within the cluster, apart from being based on teamwork activity, requires a relevant strategy and a regular review of its objectives. Thus, the cluster's ability to adjust its activities and their focus accordingly to the phase in its life cycle constitutes an important characteristic. Extensive attention in the book was dedicated to the issue of cluster activities including distinguishing cluster best practices from Poland, the Czech Republic, the Slovak Republic and Austria.

In addition, key (critical) factors conducive to successful cluster development were distinguished and closer attention was paid to their significance in accordance with the country of origin of the clusters. Thereby, the selection of these factors allowed for an increased focus not only on the most important achievements already accomplished by clusters, which may serve as examples to follow, but also constitutes a source of knowledge regarding the most serious problems encountered by clusters. Moreover, further attention was devoted to benchmarking as an up-to-date and growing in popularity tool for monitoring the performance of clusters and their actors, helping to identify weak points and consequently to adopt measures leading to necessary improvements conditioning cluster excellence.

The increasing importance of globalisation places a new perspective on the development of networking and clusters, which in order to develop optimally need excellence on the macro (policy, government) as well as on the micro level (cluster organisation). Therefore, this book aimed at analysing the existing conditions, investigating the steps implemented and identifying good practices as well as hampering factors at different levels. A profound analysis of the current stage of development of cluster-based policies and clusters themselves, with a special focus on cluster activities supporting the performance of the entities involved and the factors conducive to the successful development of clusters in the chosen countries, was conducted. On the basis of this analysis, recommendations were formulated and a proposal for a framework for cluster development in the selected countries was presented.

Additionally, the conducted research drew attention to current trends in cluster development such as the concept of inter-cluster co-operation. As clusters are acknowledged as an important tool of internationalisation, the new approach to inter-cluster co-operation with its positive (opportunities) and negative (threats) aspects being an important component for obtaining and maintaining a competitive advantage over others, was also presented. Here, in reference to the findings obtained regarding cluster development in the selected countries, the extension of co-operation among clusters with an emphasis placed on an international level of collaboration should be stimulated. The development of inter-cluster co-operation on an international level and its principles in general constitutes an interesting concept for further investigation on an international scale.

Last but not least, the findings presented in this book should be beneficial for both clusters as well as institutions supporting their development. Mapping the positive aspects as well as the emphasis placed on identifying existing imperfections allowed for the formulation of recommendations, which can be further used within the development of cluster-based policies and programmes of support for the new programming period in these countries.

SUMMARY

This book, titled "The Clusters Phenomenon in the Selected Central European Countries", presents the current research results with regard to clusters and cluster policy development, focusing on Central European countries such as Poland, the Czech Republic, the Slovak Republic and Austria.

It offers a profound analysis of the current status of both cluster-based policies and clusters' development in the selected countries (investigation of achievements and problems, good and bad practices), with a special focus on cluster activities supporting the performance of the entities involved and the factors conducive to the successful development of clusters. In connection with that, it presents the framework for the successful cluster development in the selected countries. Furthermore, it raises the current issues of internationalisation and inter-cluster co-operation with its positive (opportunities) and negative (threats) aspects, and so presents a new form of collaboration within so-called World-Class clusters.

The results of the conducted research are presented in five chapters; the sixth chapter constitutes the conclusions of the outcomes of the research conducted. The first chapter introduces the origins of the cluster concept, cluster definitions, attributes and characteristics. This is followed by the introduction of the cluster policy, its aspects and character in the selected countries, with their comparison being conducted in the second chapter. The third chapter concentrates on common cluster activities and presents practical examples of their best practices based on the research sample of clusters from Poland, the Czech Republic, the Slovak Republic and Austria. The next chapter identifies key factors conditioning successful cluster development, and in consequence, depicts the framework for the successful cluster development in the selected Central European countries. In addition, it introduces benchmarking as a useful tool for the evaluation of cluster excellence. The fifth chapter focuses on the topic of cluster internationalisation, and a new form of an inter-cluster co-operation and its features.

The book provides practitioners, policy makers and academics with a guide for successful cluster development and constitutes a source of valuable thoughts and brings a broad perspective on the cluster issues.

Its findings may be found as beneficial both by clusters as well as institutions supporting their development (e.g. government, regional authorities). The presented conclusions and recommendations can be further used for cluster policy development and shaping the programmes of support for clusters for the new programming period in these countries. Practical findings may also be beneficial in the academic sphere and assist teaching activities, for example for subjects oriented on competitiveness, companies, clusters and economic development.

APPENDIX A

Appendix A-1: List of all clusters which took part in research

	Polish clusters
1	Alternatywny Klaster Informatyczny/Alternatywny.klaster.info
2	Bydgoski Klaster Przemysłowy (Bydgoszcz Industrial Cluster)
3	Dolina Ekologicznej Żywności (Organic Food Valley)
4	Dolina Lotnicza - Stowarzyszenie Grupy Przedsiębiorców Przemysłu Lotniczego (Aviation Valley, Aviation Valley Association)
5	Dolnośląski Klaster Ekoenergetyczny EEI - Energia, Ekologia, Innowacje (Lower Silesian Eco-energetic Cluster EEI - Energy, Ecology, Innovation)
6	Gdański Klaster Budowlany (Gdańsk Construction Cluster)
7	Innowacyjny Klaster Przemysłowy "KOM-CAST" (The Eastern Casting Cluster KOM-CAST)
8	Innowacyjny Klaster Generacji i Użytkowania Energii w Mega i Nano skali
9	Innowacyjny Śląski Klaster Czystych Technologii Węglowych (Innovative Silesian Cluster of Clean Carbon Technologies)
10	Klaster "3x20" (3x20 Cluster)
11	Klaster Bioenergia dla Regionu (Bioenergy for the Region Cluster)
12	Klaster E-Południe (e-South Cluster)
13	Klaster "Life Science Kraków" (LifeScience Cluster)
14	Klaster Poligraficzny Instytutu Transferu Technologii Poligraficznych (ITTP)
15	Klaster Poligraficzno-Reklamowy w Lesznie (Leszno Printing & Advertising Cluster)
16	Klaster Restauratorów i Hotelarzy (Cluster of Restaurant and Hotel Owners)
17	Klaster Spożywczy "Naturalnie z Podlasia" (Food Cluster "Naturally from Podlasie")
18	Klaster targowo-kongresowo-usługowy "Grono Targowe Kielce" (Kielce Trade Fairs)

19	Klaster Technologii Energooszczędnych Euro-Centrum (Euro-Centre Energy Saving Technologies Cluster)
20	Klaster Turystyki i Rozwoju Regionalnego "Słońce Regionu"
21	Klaster Wołowiny
22	Klaster Zaawansowanych Technologii Przemysłu Włókienniczo - Odzieżowego
23	Klaster Zielonych Technologii (Cluster of Green Technologies)
24	Lubelskie Żagle - Regionalny Klaster w Lublinie (Lublin Sails – Regional Cluster in Lublin)
25	Łódzki Rynek Hurtowy - Zjazdowa S.A.
26	Małopolsko - Podkarpacki Klaster Czystej Energii
27	Mazowiecki Klaster Druku i Reklamy "Kolorowa Kotlina"
28	Mazowiecki Klaster ICT (Mazovia Cluster ICT)
29	Mazowiecki Klaster Innowacyjnych Technologii Fotonicznych "OPTOKLASTER" (OPTOKLASTER – Mazovian Photonic Technology Cluster)
30	Mazowiecki Klaster Lotniczy Aviation Mazovia
31	Media Klaster/Klaster Medialny (Media Cluster)
32	Nadwiślański Klaster Energii Odnawialnej i Poszanowania Energii (Vistula Renewable Energy Cluster)
33	NUTRIBIOMED Cluster
34	Podkarpacki Klaster Informatyczny (Eastern Poland IT Companies Cluster)
35	Podlaski Klaster Bielizny/Klaster Bieliźniarski (Podlachia Lingerie Cluster)
36	Polski Wschodni Klaster Medyczny
37	Pomorski Klaster Spożywczy "Żywność z Pomorza"
38	SIDE CLUSTER
39	Stowarzyszenie Wschodni Klaster Doradczo - Usługowy
40	Stowarzyszenie Zachodniopomorskie Drewno i Meble (Western Pomerania wood and furniture)
41	Śląska Sieć Wyrobów Medycznych (Medical Products Network - MedSilesia)
42	Śląski Klaster Drzewny (Silesian Wooden Cluster)
43	Świętokrzysko-Podkarpacki Klaster Budowlany "Innowator" (Świętokrzysko-Podkarpacki Construction Cluster Innowator)
44	Tarnowski Klaster Przemysłowy S.A. (The Tarnów Industrial Cluster S.A.)
45	Termomax Klaster Budownictwa Energooszczędnego (Energy-Efficient Building Cluster "TERMOMAX")

46	Warmińsko-Mazurski Klaster Razem Cieplej (Warmia and Mazury Cluster TOGETHER WARMER)
47	Wielkopolski Klaster Motoryzacyjny (Motorization Cluster of Greater Poland)
48	Wielkopolski Klaster Teleinformatyczny (Data Communications Cluster of Greater Poland)
49	Wielkopolski Klaster Zaawansowanych Technik Automatyzacji ELPROTECH (Cluster of Advanced Automation Techniques ELPROTECH in Greater Poland)
50	Wspólnota Wiedzy i Innowacji w zakresie Technik Informacyjnych i Komunikacyjnych (Knowledge and Innovation Community in the Field of Information and Communication Technologies Cluster)
51	Zachodniopomorski Klaster Chemiczny Zielona Chemia (Western Pomerania chemical cluster "Green Chemistry")
	Czech clusters
1	ATOMEX GROUP
2	Bezpečnostně technologický klastr,o.s. (Safety & Security Cluster)
3	CEITEC Cluster - bioinformatics z.s.p.o.
4	CEVTECH
5	CGC-Czech Glass Cluster, družstvo
6	Cluster Network Security Monitoring
7	CLUTEX - Klastr technické textilie (CLUTEX - Cluster of Technical Textiles)
8	CREA Hydro&Energy, o.s.
9	CzechBio
10	Czech Stone Cluster
11	Český klastr přípojných vozidel a nástaveb
12	Český nanotechnologický klastr, družstvo (Czech Nanotechnology Cluster)
13	Český pivovarský klastr
14	EKOGEN, o.s.
15	Energoklastr
16	ENVICRACK - Klastr obnovitelných zdrojů energie (Envicrack – Cluster of Alternative Energy)
17	HIT klastr
18	Hydrogen - CZ o.s.
19	IT CLUSTER
20	Jihočeský lesnicko-dřevařský klastr
21	KLACR, o.s.
22	Klastr českých nábytkářů (Cluster of Czech Cabinet Makers)

23	Klastr inovací v dopravé
24	Klastr NetPro Group z.s.p.o.
25	Klastr povrchové úpravy a.s.
26	Klastr Přesného Strojírenství Vysočina (KPSV, Precision Engineering Cluster Vysocina)
27	Knowledge management cluster (previous Stavební klastr Ostrava)
28	MedChemBio
29	Moravkoslezský automobilový klastr (MAK; Moravian-Silesian Automotive Cluster)
30	Moravskoslezský dřevařský klastr (MSDK; Moravian Timber Cluster)
31	Moravskoslezský energetický klastr (MSEK; Moravian-Silesian Energy Cluster)
32	Moravský letecký klastr, o.s. (Moravian Aerospace Cluster)
33	Nanomedic a.s.
34	Národní strojírenský klastr, o.s. (Czech Machinery Cluster)
35	Nutripol
36	Omnipack
37	Plastikářský klastr (Plastics Cluster)
38	Průmyslový klastr, z.s p.o
39	Stavební klastr Ostrava, o. s.
40	Water Treatment Alliance
	Slovak clusters
1	1.Slovenský strojársky klaster
2	Automobilový klaster - západné Slovensko (Automotive Cluster-West Slovakia)
3	Balnea Cluster Dudince
4	BITERAP
5	IKT Klaster ZAICT
6	IT Klaster Košice (Košice IT Valley)
7	Klaster cestovného ruchu-západné Slovensko
8	Klaster TURIEC
9	Slovenský Plastikársky Klaster
	Austrian clusters
1	Bau.Energie.Umwelt.Cluster Niederösterreich (Greenbuilding Cluster Lower Austria)
2	Cluster Life Sciences Tirol
3	Cluster Wellness Tirol
4	Gesundheits-Cluster (Health Technology Cluster)
5	GIS-Cluster Salzburg

6	Holzcluster Salzburg
7	Holzcluster Steiermark GmbH
8	Holzcluster Tirol
9	Initial Factor
10	Kunststoff-Cluster (Plastics-Cluster)
11	Kunststoff-Cluster Niederösterreich (Plastics Cluster Lower Austria)
12	Lebensmittel Cluster OÖ
13	Mechatronik-Cluster
14	OEC (Oekoenergie-Cluster)
15	Styrian Cleantech Cluster ECO WORLD STYRIA
16	Umwelttechnik-Cluster

List of Abbreviations and Acronyms

ABA	Austrian Business Agency
AIA	Automotive Industry Association
AOEM	Austrian Original Equipment Manufacturers
APQC	American Productivity and Quality Centre
AWS	Austrian Economic Service
BMF	Federal Ministry of Finance (Austria)
BMVIT	Federal Ministry of Transport, Innovation and Technology (Austria)
BMWA	Federal Ministry of Economics and Labour (Austria)
BMWF	Federal Ministry of Science and Research (Austria)
BSC	Balanced Scorecard
BSR InnoNet	Baltic Sea Region Innovation Network
CE	Central European
CEE-Cluster-Network	Central and Eastern European Cluster and Network Area
CI	Cluster initiative
CIP	Competitiveness and Innovation Framework Programme (2007-2013)
CIR-CE	Co-operation and Innovation and Research with Central and Eastern Europe
CLOE	Clusters Linked Over Europe
CLUNET	Cluster Network
Clusterland	Clusterland Upper Austria
CNCB	Cluster and Network Co-operation for Business Success in Central Europe
COIN	Co-operation and Networks for Innovation
COMET	Competence Centres for Excellent Technologies
CORNET	Collective Research Networking
CzechInvest	Investment and Business Development Agency
EACP	European Aerospace Cluster Partnership
ECA	European Cluster Alliance
ECEI	European Cluster Excellence Initiative
Ecoplus	Business Agency of Lower Austria

ECPG	European Cluster Policy Group
EIF	European Investment Fund
ESF	European Social Fund
ETC	European Territorial Co-operation Programme
FFG	Austrian Research Promotion Agency
FINE	Food Innovation Network Europe
FP7	Seventh Framework Programme
IBnGR	Gdańsk Institute for Market Economics
IE	Innovation Express
IFC	Institution for collaboration
INNET	Networking of national/regional funding and innovation organisations for the involvement of SMEs in technology-based innovation clusters in Europe
IROP	Integrated Regional Operational Programme (2004-2006)
ITG Salzburg	Innovation and Technology Transfer Salzburg
IWI	Institute for Industry Studies
KIS-IP	European Knowledge Intensive Services Innovation Platform
KPSV	Precision Engineering Cluster (Klastr Přesného Strojírenství Vysočina)
KWF	Carinthia Economic Promotion Fund
LISA	Life Science Austria
LISA VR	Life Science Austria Vienna Region
MIT	Ministry of Industry and Trade (Czech Republic)
MSDK	Moravian Timber Cluster (Moravskoslezský dřevařský klastr)
NADSME	National Agency for Development of SME's
NCA	National Cluster Agency
Netmatch	Innovation Network Denmark
NSRF	National Strategic Reference Framework
OECD	Organisation for Economic Co-operation and Development
OMBI	Ontario Municipal CAO's Benchmarking Initiative
OP C&EG	Operation Programme Competitiveness and Economic Growth
OP DEP	Operational Programme Development of Eastern Poland
OP EC	Operational Programme Education for Competitiveness

OP HC	Operational Programme Human Capital
OP HRE	Operational Programme Human Resources and Employment
OP R&DI	Operational Programme Research and Development for Innovation
OPEI	Operational Programme Enterprise and Innovation 2007-2013
OPIE	Operational Programme Industry and Enterprise 2004-2006
OP IE	Operational Programme Innovative Economy
PAED	Polish Agency for Enterprise Development
PI	Innovation Portal
RFTE	Austrian Council for Research and Technology Development
RIS	Regional Innovation Strategy
ROP	Regional Operational Programme
SARIO	Slovak Agency of Trade and Investments
SBA	Small Business Act
SFG	Styria Business Development Corporation
SIEA	Slovak Energy and Innovation Agency
SME	Small and Medium-sized Enterprises
TBU	Tomas Bata University
TIC	Technology Innovation Centre
TMG	Upper Technology and Marketing Corporation
UKS	Union of Slovak Clusters
UNIDO	United Nations Industrial Development Organisation
V4	Visegrad Group
WCC	World-Class Clusters
WIBAG	Burgenland Economic Service
WIFO	Austrian Centre for Economic Research
WISTO	Vorarlberg Business Location Agency
WKO	Austrian Federal Economic Chamber
WWFF	Vienna Business Agency
WWTF	Vienna Science and Technology Fund
ZIT	Centre for Innovation and Technology

BIBLIOGRAPHY

"A Practical Guide to Cluster Development. A Report to the Department of Trade and Industry and the English RDAs by Ecotec Research & Consulting", England's Regional Development Agencies, 2004, accessed May 20, 2010, http://www.dti.gov.uk/files/file14008.pdf.

American Productivity & Quality Centre (APQC). 1993, *The Benchmarking Management Guide.* Kreton: Productivity Press.

Andersen, Torsten, Bjerre Markus, and Emily Wise Hanson, 2006, *The Cluster Benchmarking Project*. Nordic Innovation Centre, accessed November 2, 2008 http://www.nordicinnovation.net/_img/cluster_benchmarking_project_final_report.pdf.

Andersen, Bjørn, and Per-Gaute Pettersen, 1996, *The Benchmarking Handbook: Step by Step Instructions*. London: Springer.

Andersson, Tomas, Schwaag-Serger, Sylvia, Sörvik, Jens, and Emily Wise Hannson, 2004, *The Cluster Policies Whitebook.* Sweden: IKED – International Organisation for Knowledge Economy and Enterprise Development.

Audretsch, David B., and Maryann P. Feldman, 1996, "Innovative Clusters and the Industry Life Cycle", *Review of Industrial Organization* 11(2): 253–273. Accessed August 22, 2011,
http://dx.doi.org/10.1007/BF00157670

Austrian Business Agency (ABA). 2008, "Cluster". frequently updated. [Vienna (Austria)]: Österreichische Industrieansiedlungs- und WirtschaftswerbungsgmbH, accessed August 22, 2011,
http://www.investinaustria.at/DE/Forschung%20%20Entwicklung/Cluster/Cluster.aspx.

Bathelt, Harald 2005, "Cluster relations in the media industry: exploring the 'distanced neighbour' paradox in Leipzig". *Regional Studies* 39(1): 105–127.

Bathelt, Harald, Malmberg Anders, and Peter Maskell 2004, "Clusters and knowledge: local buzz, global pipelines and the process of knowledge creation". *Progress in Human Geography* 28(1): 31-56.

Bergman, Edward M. 2008, "Cluster life-cycles: an emerging synthesis". In *Handbook of Research on Cluster Theory, Handbooks of research on clusters series*, edited by Charlie Karlsson, 114–132. Northampton, MA: Edward Elgar.

Bialic-Davendra, Magdalena, and Drahomíra Pavelková 2010, "Cluster concept development – a case study from Poland". Paper presented at the International Conference on Development, Energy, Environment, Economics (DEEE'10), Puerto De La Cruz, Tenerife, Spain, November 30 – December 3, 2010.

Bialic-Davendra, Magdalena 2011, "An investigation of a new concept of World-Class Clusters in Europe – A case study from Visegrad Group countries". *Journal of Competitiveness* (2): 43-57.

—. 2011, "Proposal of a framework for a cluster development in the selected Central European countries". Ph.D. diss. Tomas Bata Univeristy in Zlin.

Blien, Uwe, and Gunther Maier 2008, *The Economics of Regional Clusters. Networks, Technology and Policy.* UK: Edward Elgar Publishing Limited.

Błasiak-Nowak, Beata 2007, "Klastery jako sposób na przyspieszenie rozwoju regionów problemowych". *Gospodarka Narodowa* (7-8): 82-104.

Boschma, Ron A., and Koen Frenken 2006, "Why is economic geography not an evolutionary science? Towards an evolutionary economic geography". *Journal of Economic Geography* 6(3): 273–302. Accessed February 28, 2011, http://dx.doi.org/10.1093/jeg/lbi022

Brandenburger Adam M. and Barry J. Nalebuff 1996, *Co-Opetition.* Harper Collins Business.

Brenner, Thomas 2004, *Local Industry Cluster: Existence, Emergence and Evolution.* London and New York: Routledge.

Broderick, Anne, Garry, Tony, and Mark Beasley, 2010, "The need for adaptive processe of benchmarking in small business-to-business services". *Journal of Business & Industrial Marketing* 25(5): 324-337.

Brodzicki, Tomasz and Stanisław Szultka, 2002, "Koncepcja klastrów a konkurencyjność przedsiębiorstw". *Organizacja i Kierowanie* (4): 45-60.

"Broszura informacyjna Program Operacyjny Rozwój Polski Wschodniej". Ministerstwo Rozwoju Regionalnego, accessed April 6, 2011, http://www.polskawschodnia.gov.pl/NaborWnioskow/Documents/PORPW.pdf.

Camp, Robert C. 1989, *Benchmarking: The Search for Industry Best Practices that Lead to Superior Performance.* ASQC Quality Press, Milwaukee, WI.

—. 1995, *Business Process Benchmarking: Finding and Implementing Best Practices.* Chicago: Irwin Professional Publishing.

Carpinetti, Luiz Cesar Ribeiro, and O. T. Oiko, 2007, *Benchmarking in Clusters of SMEs: Development and Application of Database*. Paper presented at the 19th International Conference on Production Research, Chile. Accessed November 2, 2008, http://www.icpr19.cl/mswl/Papers/261.pdf.

Carpinetti, Luiz Cesar Ribeiro, and O. T. Oiko 2008, "Development and application of a benchmarking information system in clusters of SMEs". *Benchmarking: An International Journal* 15(3): 292-306.

Cassell, Catherine, Nadin, Sara. and Melanie Older Gray, 2001, "The use and effectiveness of benchmarking in SMEs". *Benchmarking: An International Journal* 8(3): 212-22.

Cassidy, Erin, Davis, Charles H., Arthurs, David, and David Wolfe, 2005, *Measuring the National Research Council's technology cluster initiatives.* Accessed November 2, 2008, http://www.cric.com.au/ cric_cluster/ cc_resources/ Cassidy-refereed.pdf.

"Cíle Národní klastrové strategie 2005 – 2008". Vláda České Republiky. Accessed February 28, 2011, http://www.mpo.cz/dokument6216.html

Clement, Werner 2010, *Beiträge zu: Leitlinien für eine österreichische Clusterstrategie*, 1st ed. [Vienna (Austria)]: Bundesministerium für Wirtschaft und Arbeit, accessed August 4, 2011, http://www.clusterplattform.at/fileadmin/user_upload/clusterbibliothek /Clement_Studie_Endversion_Clusterleitlinien__30_12_2010.pdf.

Clement, Werner and Silvia Welbich-Macek, 2007, *Erfolgsgeschichte: 15 Jahre Clusterinitiativen in Österreich*. Endbericht im Auftrag des BMWA. 1st ed. [Vienna (Austria)]: Bundesministerium für Wirtschaft und Arbeit, 4.4.4: Ökobau Cluster NÖ. Accessed August 4, 2011, http://www.clusterplattform.at/fileadmin/user_upload/clusterbibliothek /75_Erfolgsgeschichte_15_Jahre_Clusterinitiativen_in_OE.pdf.

"Cluster Management Guide – Guidelines for the Development and Management of Cluster Initiatives", CLOE – Clusters Linked Over Europe, 2006, accessed May 31, 2010, http://www.clusterforum.org/

Clusterplattform Österreich. *clusterplattform.at*. Frequently updated. [Linz (Austria)]: Bundesministerium für Wirtschaft, Familie und Jugend, August 2011, accessed August 22, 2011, http://www.clusterplattform.at.

"Cluster policy in Europe. A brief summary of cluster policies in 31 European countries", Europe Innova Cluster Mapping Project, 2008, accessed January 20, 2009, http://www.clusterobservatory.eu/upload/ Synthesis_report_cluster_mapping.pdf.

Cooke, Philip 2002, *Knowledge Economies: Clusters, Learning and Co-operative Advantage.* London: Routledge.

Council Decision of 6 October 2006 on Community strategic guidelines on cohesion (2006/702/EC). Official Journal of the European Union. L 291/11, 21.10.2006, accessed February 27, 2011, http://ec.europa.eu/regional_policy/sources/docoffic/2007/osc/l_29120061021en00110032.pdf.

Czech Society for Quality (Česká společnost pro jakost). 2002, *Benchmarking – Text pro účastníky*. 1. ed.

CzechInvest. 2005, *Průvodce klastrem*. Accessed August 12, 2011, http://www.czechinvest.org/data/files/pruvodce-klastrem-63.pdf.

Dalum, Bent, Pedersen, Christian O.R., and Gert Villumsen 2005, "Technological life-cycles; lessons from a cluster facing disruption". *European Urban and Regional Studies* 12(3): 229-46. Accessed February 28, 2011, http://dx.doi.org/10.1177/0969776405056594

Davies, A., and A. Kochbar, 2002, "Manufacturing best practice and performance studies: a critique", *International Journal of Operations & Production Management* 22(3): 289-305.

Decision No.1639/2006/EC of the European Parliament and of the Council of 24 October 2006 establishing a Competitiveness and Innovation Framework Programme (2007-2013). Accessed February 27, 2011, http://europa.eu/legislation_summaries/energy/european_energy_policy/n26104_en.htm.

Department of Trade and Industry. 2003. *A Practical Guide to Cluster Development*. Accessed October 28, 2008, http://www.dti.gov.uk/clusters.

Dervojeda, Kristina 2011, *Competitiveness and Cluster Performance Measurement Systems Survival of the Adaptive*. Paper presented at the 14th TCI Annual Global Conference, Auckland, New Zealand.

Enright, Michael J. 1998, The *Globalization of Competition and the Localization of Competitive Advantage: Policies toward Regional Clustering*. Paper presented at the Workshop on the Globalization of Multinational Enterprise Activity and Economic Development University of Strathclyde Glasgow, Scotland, May 1998.

—. 2003, "Regional Clusters: What We Know and What We Should Know". In *Innovation Clusters and Interregional Competition*, edited by Johannes Bröcker, Dirk Dohse and Rüdiger Soltwedel, 99-129. Berlin: Springer-Velrag.

Etzkowitz, Henry 2002, The *Triple Helix of University-Industry-Government, Implications for Policy and Evaluation*. Accessed January 20, 2009, http://www.sister.nu/pdf/wp_11.pdf.

Europa InterCluster. 2010, *White Paper. The Emerging of European World-Class Clusters*. Belgium, accessed February 8, 2011, http://www.intercluster.eu/images/stories/white_paper/white_paper_the_emerging_of_european_world_class_clusters.pdf.

"European Cluster Memorandum", accessed December 7, 2010, http://www.proinno-europe.eu/NWEV/uploaded_documents/European_Cluster_Memorandum.pdf.

European Cluster Policy Group. 2010, *Consolidated Set of Policy Recommendations on Four Themes.* Accessed August 22, 2011, http://www.proinno-europe.eu/ecpg/newsroom/ecpg-final-recommendations.

European Commission. 2005, *Final report of the Expert Group on Enterprise Clusters and Networks.* European Commission, Belgium.

—. *Putting knowledge into practice: A broad-based innovation strategy for the EU.* COM(2006) 502 final. Accessed July 12, 2010, http://eur-lex.europa.eu/LexUriServ/site/en/com/2006/com2006_0502en01.pdf.

—. Commission Communication "*Towards world-class clusters in the European Union: Implementing the broad-based innovation strategy*". COM(2008) 352 final.

—. *EUROPE 2020. A European strategy for smart, sustainable and inclusive growth*. Accessed December 7, 2010, http://ec.europa.eu/eu2020/index_en.htm.

European Union Parliament. Lisbon European Council 23 and 24 March Presidency Conclusion. Accessed December 7, 2010, http://www.europarl.europa.eu/summits/lis1_en.htm.

Evers, Hans-Dieter 2008, *Knowledge hubs and knowledge clusters: Designing a knowledge architecture for development.* Centre for Development Research (ZEF), University of Bonn, accessed April 5, 2010, http://mpra.ub.uni-muenchen.de/8778/.

Furre, Harald 2007, *Country Report: Austria,* accessed December 19, 2011, http://www.clusterobservatory.eu/index.html#!view=documents;mode=one;sort=name;uid=8e887503-e1ff-4222-98fd-a386558049a8;id=.

Gomes, Carlos F., and Mahmoud M. Yasin, 2011, "A systematic benchmarking perspective on performance management of global small to medium-sized organisations: An implementation-based approach", *Benchmarking: An International Journal* 18(4): 543-562.

Gorynia, Marian, and Barbara Jankowska, 2007, "Wpływ klasterów na konkurencyjność i internacjonalizację przedsiębiorstw", *Gospodarka Narodowa* (7-8): 1-18.

Gorynia, Marian, and Barbara Jankowska, 2008, *Klastry a międzynarodowa konkurencyjność i internacjonalizacja przedsiębiorstwa.* Warszawa: Difin.

Górzyński, Michał 2006, "System wsparcia gron w województwie podkarpackim – rekomendacje". In *System wspierania gron przedsiębiorczości – publikacja podsumowująca,* edited by Michał Górzyński, 95-119. Rzeszów: IG WSIiZ.

Grabher, Gernot 1993, "The weakness of strong ties: the "lock-in" of regional development in the Ruhr area", In *The Embedded Firm: On the Socio-Economics of Industrial Networks*, edited by Gernot Grabher. London: Routledge.

Grycuk, Adrian 2003, "Koncepcja gron w teorii i praktyce zarządzania", *Organizacja i Kierowanie* 3(113): 3-16.

"Guide Book. 13 ways to become innovative. Establishing a successful business within the Operational Programme Innovative Economy", 2009. Warsaw: PAED.

Gulda, Krzysztof 2008, Klastry jako szansa dla przedsiębiorczości. In *Wspieranie rozwoju klastrów w Polsce i zagranicą – doświadczenia i wyzwania*, edited by Maciej Dzierżanowski and Stanisław Szultka, 8-9. Gdańsk: IBnGR.

Hassink, Robert 2005, "How to Unlock Regional Economies from Path Dependency? From Learning Region to Learning Cluster", *European Planning Studies* 13(4): 521-535.
http://dx.doi.org/10.1080/09654310500107134

Hegedüs, V. 2008, "Theoretical approach to clusters and applied cluster policy", *Development and Finance: Quarterly Hungarian Economic Review* (2): 77-87.

Hendry, Chris, Brown, James, and Robert Defillippi, 2000, "Regional clustering of high technology-based firms: opto-electronics in three countries", *Regional studies* (34): 129-144.

Hutschenreiter, Gernot 1994, *Cluster innovativer Aktivitäten in der österreichischen Wirtschaft*. TIP, Vienna.

Hwang, L.-J., and Andrew Lockwood, 2006, "Understanding the challenges of implementing best practices in hospitality and tourism SMEs", *Benchmarking: An International Journal* 13(3): 337-54.

Innovating Regions in Europe. Accessed February 7, 2009, http://www.innovating-regions.org/download/Clusters_in_CCE.pdf

—. "IRE subgroup on Regional clustering and networking as innovation drivers: Cluster Management (Learning module 5)". Accessed May 5, 2011, www.innovating-regions.org.

—. "IRE subgroup on Regional clustering and networking as innovation drivers: The future of cluster policies and programmes – New demands and their implications (Learning module 6)". Accessed May 5, 2011, www.innovating-regions.org.

"Innovation Clusters in Europe: A statistical analysis and overview of current policy support". DG Enterprise and Industry Report, accessed February 27, 2011, http://www.proinno-europe.eu/admin/uploaded_documents/innovation_clusters_in_europe.pdf.

Institute for Strategy and Competitiveness. 2002, "The Cluster Meta-Study". Accessed October 20, 2008, http://data.isc.hbs.edu/ MetaStudy2002Bib.pdf.

—. 2003, "The Cluster Mapping Project". Accessed October 20, 2008, http://data.isc.hbs.edu isc/index.jsp.

"Instrumenty wspierania klastrów", accessed May 30, 2011, http://www.mg.gov.pl/Wspieranie+przedsiebiorczosci/Wsparcie+finansowe+i+inwestycjeKlastry/Instrumenty+wspierania+klastrow.

Integrated Guidelines for Growth and Jobs (2005-2008). Commission of EC. Brussels, 12.4.2005. COM(2005) 141 final.

Jagódka, Maciej 2009, "Istota, zasady funkcjonowania i znaczenie klastrów dla rozwoju przedsiębiorstw", *Nauka i Gospodarka* (2): 28-36.

Jirčíková, Eva 2008, "Identification of Factors Influencing the Cluster Development Process in the Czech Republic". Ph.D. diss. Tomas Bata Univeristy in Zlin.

Kačirková, Mária 2008, "Zhlukový potenciál v regiónoch nových členských krajín Európskiej Únie", *Ekonomickě Rozhľady* 4: 448-461.

Kaplan, Robert S. 1983, "Measuring Manufacturing Performance: A New Challenge for Managerial Accounting Research", *Accounting Review* 58(4): 686-705.

Karlöf, Bengt, and Svante Östblom 1995, *Benchmarking: Jak napodobit úspěšné.* 1. ed. Praha: Victoria Publishing.

Ketels, Christian 2009, *Clusters, Cluster Policy, and Swedish Competitiveness in the Global Economy. Expert Report no. 30 to Sweden's Globalisation Council.* Västerås: PRINT Edita.

Ketels, Christian, Lindqvist, Göran, and Örjan Sőlvell 2007, *Cluster and Cluster organizations.* Centre for Strategy and Competitiveness, Stockholm School of Economics.

Ketels, Christian, Lindqvist, Göran, and Örjan Sőlvell 2008, *Cluster and Cluster Initiatives*. Centre for Strategy and Competitiveness, Stockholm School of Economics, accessed May 7, 2011, http://www.europe-innova.eu/c/document_library/get_file?folderId=148901&name=DLFE-9310.pdf.

Ketels, Christian, and Olga Memedovic 2008, From clusters to cluster-based economic development, *Int. J. Technological Learning, Innovation and Development* 1(3): 375–392.

"Kierunki zwiększania innowacyjności gospodarki na lata 2007-2013", accessed February 7, 2009, http://www.mg.gov.pl/NR/rdonlyres/90AF42C4-A420-4BF9-9CE8-08C28B8E4FFE/50360/KierunkiPL.pdf.

Klepper, Steven 1996, "Entry, exit, growth and innovation over the product life cycle", *American Economic Review* 86(3): 562-583.

—. 1997, "Industry life cycles", *Industrial and Corporate Change* 6(1): 145-81.

Kosińska, Sylwia 2007, *Country Report: Poland,* accessed January 5, 2011, http://www.clusterobservatory.eu/index.php?id=42&nid=.

Kryjom, Piotr 2008, "Doświadczenia PARP z wdrażania instrumentów wspierających powiązania kooperacyjne (klastry)". In *Wspieranie rozwoju klastrów w Polsce i zagranicą – doświadczenia i wyzwania,* edited by Maciej Dzierżanowski and Stanisław Szultka, 20-23. Gdańsk: IBnGR.

"Kształtowanie polityki klastrowej w Polsce", accessed May 30, 2011, http://www.mg.gov.pl/node/11789.

Lehner, Patrick, Bergman, Edward, and Gunther Maier 1998, Business Clusters in der österreichischen Wirtschaft. *IMONET zwischenbericht.* IIR, Wien.

Lorenzen, Mark 2005, "Why do clusters change?", *European Urban and Regional Studies* 12(3): 203–208. http://dx.doi.org/10.1177/0969776405059046

Malakauskaite, Asta, and Valentinas Navickas 2010, "Relation between the Level of Clusterization and Tourism Sector Competitiveness", *Inzinerine Ekonomika- Engineering Economics* 21(1): 60-67.

Malakauskaite, Asta, and Valentinas Navickas 2011, "Relation between a Cluster's Life Cycle and Competitiveness Level", *Economics and Management (Ekonomika ir Vadyba)* 16: 214-218.

"Małopolski Regionalny Program Operacyjny na lata 2007-2013". Zarząd Województwa Małopolskiego, accessed February 28, 2011, http://www.malopolskie.pl/Pliki/2007/MRPO_20071004.pdf.

Marshall, Alfred 1920, *Principles of Economics.* 8th ed. London: Macmillan.

Martin, Ron, and Peter Sunley 2006, "Path dependence and regional economic evolution", *Journal of Economic Geography* 6(4): 395–437. http://dx.doi.org/10.1093/jeg/lbl012

Maskell, Peter, and Leila Kebir 2005, "What qualifies as a cluster theory?" DRUID Working Paper No.05-09. Department of Industrial Economics and Strategy, Copenhagen Business School.

McAdam, Rodney, and Edel O´Neill 1999, "Taking a critical perspective to the European business excellence model using a balanced scorecard approach: a case study in the service sector", *Manging Service Quality* 9(3): 191-7.

McCarthy, Breda 2003, "Strategy is personality-driven, strategy is crisis driven: insights from entrepreneurial firms", *Management Decision* 41(4): 327-39.

McKay, Ruth, and E. Chung 2005, "Benchmarking for entrepreneurial survival", *Benchmarking: An International Journal* 12(3): 207-16.

"Memorandum of Understanding on Co-operation in the field of the Cluster Development of the V4 Countries", Budapest, 26 November 2009. Accessed December 7, 2010, http://www.arr.cz/userfiles/file/Klastry/MOU_on_V4_Cluster_Co-operation.pdf.

Menzel, Max-Peter, and Dirk Fornahl 2009, "Cluster life cycles-dimensions and rationales of cluster evolution", *Industrial and Corporate Change* 19(1): 205–238. http://dx.doi.org/10.1093/icc/dtp036

Mikołajczak, Bożena, Kruszewska, Agnieszka, and Joanna Fila 2009, *Klastry na świecie. Studia przypadków*. Warszawa: Difin.

Moriarty, John P. 2011, "A theory of benchmarking", *Benchmarking: An International Journal* 18(4): 588-608.

Mytelka, Lynn, and Fulvia Farinelli 2000, "Local Clusters, Innovation Systems and Sustained Competitiveness". The United Nations University. Discussion Paper Series. Accessed May 5, 2009, http://www.intech.unu.edu/publications/discussion-papers/2000-5.pdf.

"Národný strategický referenčný rámec 2007 – 2013". Ministerstvo Výstavby a Regionálneho Rozvoja Slovenskej Republiky, accessed February 28, 2011, http://www.nsrr.sk/narodny-strategicky-referencny-ramec-2007-2013/.

Neužilová Iveta "Národní klastrová strategie 2005 – 2008". Accessed January 17, 2012, http://www.mpo.cz/dokument6216.html.

O'Callaghan Ramon 2007, "On Dynamic Clustering, Innovation, and the Role of IT". *In The Creative Enterprise: Managing Innovative Organizations and People*, edited by Davila, Tony, Epstein, Marc J. and Robert Shelton, 17-42, USA: Praeger Publishers.

OECD. 1999, *Boosting innovation. The Cluster Approach*. Paris: OECD Publishing.

—. "LEED-CEI-EBRD project on Industrial Districts and Local Clusters 2001-2002". Accessed February 27, 2011, http://www.oecd.org/document/6/0,3343,en_2649_34457_2438406_1_1_1_1,00.html.

Ontario Municipal CAO's Benchmarking Initiative (OMBI). 2001. *Creating a Municipal Performance Measurement Framework.* Accessed March 7, 2009, http://www.ombi.ca/docs/newsinfo.asp?track =vi&itemid=26.

"Operační Program Průmysl a Podnikání na léta 2004 – 2006". Ministerstvo Průmyslu a Obchodu České Republiky, accessed February 28, 2011, http://www.evaluace.cz/dokumenty/oper_progr_cr/ op_prumysl.pdf.

"Operačný Program Zamestnanosť a Sociálna Inklúzia". Ministerstvo Práce, Sociálnych Vecí a Rodiny Slovenskej Republiky, accessed February 28, 2011, http://www.esf.gov.sk/documents/OP2007/ OpZaSI_Final2007.pdf.

"Operational Programme Competitiveness and Economic Growth". Ministry of Economy of the Slovak Republic, accessed February 28, 2011, http://www.nsrr.sk/en/operational-programmes/competitiveness-and-economic-growth/.

"Operational Programme Development of Eastern Poland 2007 – 2013". Warsaw: Ministry of Regional Development, October 2, 2007. Accessed February 28, 2011, http://www.mrr.gov.pl/english/ european_funds_2007_2013/european_funds_2007_2013/PORPW/Do cuments/PORPW_eng_final.pdf.

"Operational Programme Education for Competitiveness". Ministry of Education, Youth and Sports, accessed May 23, 2011, http://www.strukturalni-fondy.cz/getdoc/5257c95e-fac7-45bb-8313-a489678b9010/Dokumenty.

"Operational Programme Enterprise and Innovation". Ministry of Industry and Trade of the Czech Republic, accessed February 28, 2011, http://www.czechinvest.org/data/files/oppi-msc-en-29-11-schvalen-ek-674.pdf.

"Operational Programme Human Resources and Employment 2007-2013". Ministry of Labour and Social Affairs, accessed February 28, 2011, http://www.strukturalni-fondy.cz/getdoc/61e1c103-49f7-4d5a-b809-1142bce99472/Dokumenty.

"Operational Programme Research and Development for Innovation". Ministry of Education, Youth and Sports, July 24, 2008. Accessed May 23, 2011, http://www.msmt.cz/strukturalni-fondy/operacni-program-vyzkum-a-vyvoj-pro-inovace-1.

Quinn, M. 2009, Reviewing innovation: Changing conditions and new challenges. Presented at the 12th TCI Annual Global Conference, Jyväskylä, Finland, October 2009.

PARP. 2010, *Benchmarking klastrów w Polsce – 2010. Raport z badania.* Warszawa.

—. 2012, *Benchmarking klastrów w Polsce – edycja 2012. Raport z badania.* Warszawa.

Pavelková, Drahomíra, Bialic-Davendra, Magdalena, Jirčíková, Eva and Lubor Homolka 2013, "Clusters' Activities and Economy Stage of Development: Evidence from V4 and Advanced Economies", *Ekonomický časopis (Journal of Economics)* 61(2): 187 – 205.

Pavelková, Drahomíra, Friedel, Libor., Jirčiková, Eva, Knápková, Adriana, Skokan, Karel, and Petra Škodáková 2009, *Klastry a jejich vliv na výkonnost firem.* Prague: GRADA.

Peneder, Michael 1994, *Clusteranalyse und sektorale Wettbewerbsfähigkeit der österreichischen Industrie.* WIFO, Vienna.

Pilarska, Czesława 2009, "Rola państwa w procesie tworzenia i rozwoju struktur klastrowych w Polsce", *Prace Naukowe Uniwersytetu Ekonomicznego we Wrocławiu* (39): 799-815.

—. 2010, "Polityka oparta na klastrach w wybranych krajach Unii Europejskiej", *Zeszyty Naukowe Polskiego Towarzystwa Ekonomicznego* (8): 89-118.

Poledna, Juraj 2007, *Country Report: Czech Republic.* Accessed October 26, 2009, http://www.clusterobservatory.eu/index.php?id=42&nid=.

Porter, Michael E. 1990, *The Competitive Advantage of Nations.* 1st ed. New York: Free Press.

—. 1998, "Clusters and the New Economics of Competition", *Harvard Business Review.* November-December, 77-90.

—. 2008, *On Competition.* Updated and Expanded Edition. Boston: Harvard Business School Publishing.

Pouder, Richard, and Caron H. St. John 1996, "Hot spots and blind spots: geographical clusters of firms and innovation", *Academy of Management Review* 21(4): 1192-225. http://dx.doi.org/10.5465/AMR.1996.9704071867

Presidency Conclusions of the Brussels European Council (13-14 March 2008). Accessed December 8, 2010, http://www.consilium.europa.eu/ueDocs/cms_Data/docs/pressData/en/ec/99410.pdf.

PRO-INNO Europe. "CLUNET Policy Guidelines Report". Accessed December 8, 2010, http://observatoirepc.org/fileadmin/user_upload/Ressources/CLUNET_policy_Guidelines_080108.pdf.

"Przewodnik po kryteriach wyboru finansowanych operacji w ramach Programu Operacyjnego Innowacyjna Gospodarka, 2007-2013", 2008. Warszawa: Ministerstwo Rozwoju Regionalnego. Accessed May 5,

2011, http://www.konkurencyjnosc.gov.pl/NR/rdonlyres/2DBE47EF-384E-4986-9E5C3FDE44DBCF66/46624/Przewodnik_po_kryteriach_200509.pdf.

Rat Für Forschung Und Technologieentwicklung (RFTE). 2009. *Empfehlung zur österreichischen Clusterinitative*. 1st ed. [Vienna (Austria)]: RFTE, accessed August 4, 2011, http://www.rat-fte.at/tl_files/uploads/Empfehlungen/090512_Clusterempfehlung.pdf.

"Regionálna inovačná stratégia Bratislavského samosprávneho kraja". 2004. Záverečná správa v rámci projektu RIS SK. BIC Bratislava, accessed May 5, 2011, http://www.region-bsk.sk/SCRIPT/ViewFile.aspx?docid=56.

"Regionálna inovačná stratégia Košického samosprávneho kraja". Košický Samosprávny Kraj, accessed February 5, 2011, http://www.vucke.sk/APIR/sk/Pre_Podnikatelov/Investicne_prostredie/regionalnainovacnastrategia/Documents/RIS.pdf.

"Regionálna inovačná stratégia". Nitriansky Samosprávný Kraj. Accessed May 6, 2011, http://www.unsk.sk/files/regionalny/rissk_broz_sk.pdf.

"Regionálna inovačná stratégia Prešovského samosprávneho kraja". Accessed May 5, 2011, http://www.po-kraj.sk/sk/samosprava/urad/odbor-regionalneho-rozvoja-uzemneho-planu-zivotneho-prostredia/dokumenty-regionalneho-rozvoja/projekt-ris.html.

"Regionálna inovačná stratégia pre Trenčiansky samosprávny kraj". RIS INSTIT, accessed May 6, 2011, http://www.tsk.sk/buxus/docs//regional/ris/RIS_INSTIT_TSK.pdf.

"Regionálna inovačná stratégia Trnavského samosprávneho kraja". Accessed May 5, 2011, http://www.trnava-vuc.sk/index.php/dokumenty_rr_sap.html?file=tl_files/dokumenty/dokumenty_odbor_strategie_a_programovania/dokumenty/ttsk/07-RIS_TRNAVA.pdf.

"Regionální inovační strategie Jihočeského kraje". Jihočeská agentura pro podporu inovačního podnikání, accessed May 5, 2011, http://www.kraj-jihocesky.eu/index.php?par[id_v]=1545&par[lang]=CS.

"Regionální inovační strategie Jihomoravského kraje". Jihomoravské inovační centrum, accessed May 5, 2011, http://www.kr-jihomoravsky.cz/Default.aspx?pubid=234&TypeID=7&foldid=2928&foldtype=7.

"Regionální inovační strategie Královéhradeckého Kraje". Centrum EP - Regionální rozvojová agentura, accessed April 5, 2011, http://www.kr-kralovehradecky.cz/assets/rozvoj-kraje/RVVI/Strategie-_rijen09.pdf.

"Regionální inovační strategie Libereckého kraje". ARR - Agentura regionálního rozvoje, accessed May 5, 2011, http://www.portal-inovace.cz/userdata/files/RISLK_plna_verze.pdf.

"Regionální Inovační Strategie Moravskoslezského kraje 2010 – 2016". Agentura pro regionální rozvoj, accessed May 5, 2011, http://verejna-sprava.kr-moravskoslezsky.cz/assets/rozvoj_kraje/rk_ris_2010-16.pdf.

"Regionální inovační strategie Pardubického kraje". Regionální rozvojová agentura Pardubického kraje, accessed May 5, 2011, http://www.pardubickykraj.cz/article.asp?thema=4011&item=55061.

"Regionální inovační strategie Zlínského kraje", 2007. Accessed May 5, 2011, http://ris.utb.cz/.

"Regionalny Program Operacyjny Warmia i Mazury na lata 2007-2013". Zarząd Województwa Warmińsko – Mazurskiego, accessed February 28, 2011, http://www.rpo.warmia.mazury.pl/index.php?page=dzial&dzial_id=79.

"Regionalny Program Operacyjny Województwa Kujawsko-Pomorskiego na lata 2007-2013". Zarząd Województwa Kujawsko-Pomorskiego, accessed February 28, 2011, http://www.kujawsko-pomorskie.pl/files/roz_reg/prog_oper/rpo/RPO_2007-2013_WERSJA_IV.pdf.

"Regionalny Program Operacyjny Województwa Mazowieckiego 2007-2013". Urząd Marszałkowski Województwa Mazowieckiego, accessed February 28, 2011, http://rpo.mazowia.eu/data/other/rpo_1.pdf.

Rosenfeld, Stuart A. 1997, "Bringing Business Clusters into the Mainstream of Economic Development", *European Planning Studies* 5(1): 3-23.

"Rozporządzenie Ministra Gospodarki z dnia 2 grudnia 2006 r. w sprawie udzielania przez Polską Agencję Rozwoju Przedsiębiorczości pomocy finansowej niezwiązanej z programami operacyjnymi". Accessed May 7, 2011, http://isap.sejm.gov.pl/DetailsServlet?id=WDU20062261651.

Rugman, Alan M., and Alain Verbeke 2003, "Multinational Enterprises and Clusters: An Organizing Framework", *Management International Review* (43): 9-22.

Saxenian, AnnaLee 1994, *Regional Advantage: Culture and Competition in Silicon Valley and Route 128.* Cambridge: Harvard University Press.

Skawińska, Eulalia, and Romuald Zalewski 2009, *Klastry biznesowe w rowoju konkurencyjności i innowacyjności regionów. Świat-Europa-Polska.* PWE: Warszawa.

Sonderegger, Petra, and Florian Täube 2010, "Cluster life cycle and diaspora effects: Evidence from the Indian IT cluster in Bangalore", *Journal of International Management* (16): 383–397.

Sölvell, Örjan 2009, *Clusters. Balancing Evolutionary and Constructive Forces*. Second edition. Sweden: Ivory Tower.

Sölvell, Örjan, Lindqvist, Göran and Christian Ketels 2003, *The Cluster Initiative Greenbook*. Stockholm: Ivory Tower AB.

Stachowicz, Jan 2006, "Zarządzanie kapitałem społcznym procesem organizowania i rozwoju klastrów, doświadczenia z prac nad organizowaniem klastrów przemysłowych w województwie śląskim". In *Klastry jako narzędzia lokalnego i regionalnego rozwoju gospodarczego,* edited by Eva Bojar, 26-47. Lublin: Politechnika Lubelska.

Staszewska, Jolanta 2007, "Rozwój przedsiębiorstw sieciowych w Polsce – perspektywy dla klasteringu", *Przegląd Organizacji* 11: 30-32.

—. 2009, "Wykorzystanie nowoczesnej koncepcji zarzązania przez sieć – przyczyny i skutki klasteringu", *Przegląd Organizacji* 1: 29-34.

Swann, Peter G.M. 1998, "Towards a model of clustering in high-technology industries". In *The Dynamics of Industrial Clustering*, edited by Swann, Peter G.M., Prevezer, Martha, and David K. Stout, 52-76. Oxford: Oxford University Press.

Szanyi, Miklós, Akbar, Yusaf H., Buzás, Sándor, Pogátsa Zoltán and Magdolna Sass 2007, "Clusters and cluster policies: What are they, why are they important, how to promote them?" CEE-ClusterNetwork, Work package 1 (WP 1). Accessed May 5, 2012, http://www.clustercollaboration.eu/documents/10147/16786/d1.3_report_final.pdf

Szultka, Stanisław 2008, "Polityka wspierania klastrów w Polsce – doświadczenia, ocena i wnioski na przyszłość". In *Wspieranie rozwoju klastrów w Polsce i zagranicą – doświadczenia i wyzwania,* edited by Maciej Dzierżanowski, and Stanisław Szultka, 12-15. Gdańsk: IBnGR.

Szymoniuk, Barbara 2005, "Podwyższanie konkurencyjności regionu rolniczego: klaster Dolina Ekoligicznej Żywności". In *Kreowanie konkurencyjności regionu. Grona przemysłowe w regionie*, edited by Zbigniew Olesiński, and Agnieszka Predygier, Kielce: Wydział Zarządzania i Administracji Akademii Świętokrzyskiej.

Škodáková, Petra 2009, "Proposed Model for Measuring and Managing Company Performance by Benchmarking under Cluster Conditions ". Ph.D. diss. Tomas Bata Univeristy in Zlin.

Škodáková, Petra, Pavelková, Drahomíra, and Tomaš Výmola 2008, "ICT Application for Benchmarking of Financial Performance of Clusters". Centre for investigations into Information Systems (CVIS). Accessed December 1, 2008, http://www.cvis.cz/eng/.

"The NGPExcellence Project. Cluster Excellence in the Nordic Countries, Germany and Poland", accessed May 5, 2011, http://www.pi.gov.pl/PARPFiles/file/klastry/Leaflet_NGPE.pdf.
Tichy, Gunther 1998, Clusters: less dispensable and more risky than ever. In *Clusters and Regional Specialization: On Geography, Technology and Networks*, edited by Michael Steiner, 226-237. London: Pion.
Ter Wal, Anne L.J., and Ron Boschma 2011, "Co-evolution of Firms, Industries and Networks in Space", *Regional Studies* 45(7): 919-933.
Tödtling, Franz 2001, "Industrial Clusters and Cluster Policies in Austrian Regions". In *Cluster Policies – Cluster Development?*, edited by Åge Mariussen, 59-78. Stockholm: Nordregio.
UNIDO. 2001. *Development of Cluster and Networks of SMEs. The UNIDO Programme.* Vienna: UNIDO.
Utterback, James M., and William J. Abernathy 1975, "A dynamic model of process and product innovation", *Omega 3* 3(6): 639-656.
Watson, Gregory H. 1993, *Strategic Benchmarking: How to Rate Your Company's Performance against the World's Best.* New York: John Wiley & Sons, Inc.
—. 2007, *Strategic Benchmarking Reloaded with Six Sigma.* Hoboken, New Jersey: John Wiley & Sons, Inc.
Weiss, Andreas 1994, *Österreich als Standort international competitiver Cluster.* IWI Studien, Band XIII, Wien.
Wieczorek, Paweł 2008, "Klastry a pozycja rynkowa przedsiębiorców". *Ekonomika i Organizacja Przedsiębiorstwa* (12): 61-70.
Wolter, Kerstin 2003, "Life cycle for clusters? The dynamics governing regional agglomerations". Presented at the Clusters, Industrial Districts and Firms: The Challenge of Globalization, Modena, Italy.
Wójcik-Augustyniak, Marzena 2009, "Klaster edukacyjny w Siedlcach", *Zeszyty Naukowe Akademii Podlaskiej, Seria: Administracja i Zarządzanie* (83): 97-109.
Van Klink, Arjen, and Peter De Langen 2001, "Cycles in industrial clusters: the case of the shipbuilding industry in the northern Netherlands", *Tjidschrift voor Economische en Sociale Geografie* 92(4): 449-463. http://dx.doi.org/10.1111/1467-9663.00171
Žižka, Miroslav 2004, "Metody identifikace klastrů", *E+M Ekonomie a Management* 7(4): 32-46.

Used websites

http://www.autoklaster.sk/
http://clusterscord.eu/

http://www.cluster-excellence.eu/3565.html
http://www.clusterforum.org/en/about_cloe.html
http://www.clusterobservatory.eu/index.html#!view=aboutobservatory;url=/about-observatory/
http://www.clusterplattform.at/
http://www.clusnet.eu/
http://clustrat.eu/
http://www.cncb.eu/
http://cordis.europa.eu/fp7/capacities/regions-knowledge_en.html
http://www.creacz.com/
http://www.dolinalotnicza.pl/
http://ec.europa.eu/enterprise/newsroom/cf/itemlongdetail.cfm?item_id=2834
http://www.eco.at/cms/5/Deutsch/
http://www.ecoplus.at/en
http://www.elektroklastr.cz/
http://www.funduszestrukturalne.gov.pl/NR/rdonlyres/718458E5-53A5-4486-9CE2-657036614EBE/0/irop_pro.pdf
http://www.holzcluster-tirol.at/
http://klaster.info/index.php/main/pl/ocategory
http://www.klastry-efs.pl/index.php?option=com_content&task=view&id=20&Itemid=40
http://www.kpsv.cz/
http://lifescience.pl/
http://www.msdk.cz/#
http://www.nca.cz/en/nca
http://www.netmatch.nu/
http://www.pi.gov.pl/PARP/data/klastry/index.html
http://www.plastr.cz/
http://www.proinno-europe.eu/eca
http://www.uks.sk/ciele_uks.php

AUTHORS

Magdalena Bialic-Davendra
Scientific Staff at the Centre for Applied Economic Research, Faculty of Management and Economics, at Tomas Bata University in Zlín, Czech Republic.

Eva Vejmělková
Scientific Staff at the Centre for Applied Economic Research, Faculty of Management and Economics, at Tomas Bata University in Zlín, Czech Republic.

Drahomíra Pavelková
Full Professor and the Dean of the Faculty of Management and Economics, at Tomas Bata University in Zlín, Czech Republic.

INDEX